NEW ZEALAND SPIDERS

NEW ZEALAND SPIDERS

AN INTRODUCTION

with 132 colour photographs
and 164 drawings and photographs

by

R. R. FORSTER

and

L. M. FORSTER

Collins

AUCKLAND AND LONDON

COMPARATIVE MEASUREMENT SCALE

The metric system is used in scientific measurements. These two scales of centimetres – millimetres and inches – 1/12 and 1/48 inch will enable the reader to compare the measurements.

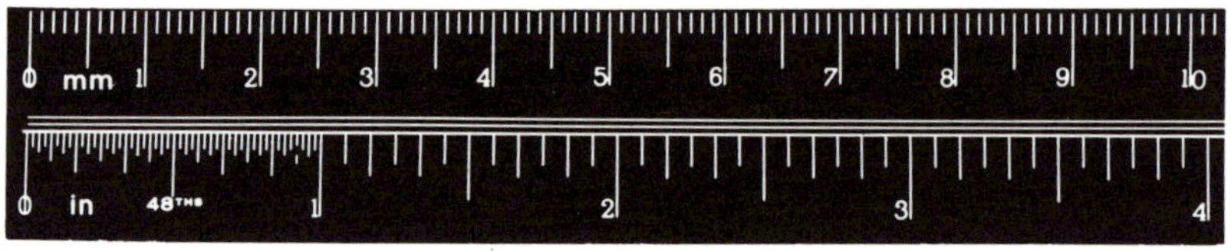

FIRST PUBLISHED 1973
COLLINS BROS. & CO. LTD.
P. O. BOX No. 1, AUCKLAND

Printed in Hong Kong by
Dai Nippon Printing Co., (International) Ltd.

Contents

Illustrations

IN COLOUR

New Zealand Spiders

BLACK AND WHITE ILLUSTRATIONS

Figure

Preface

It may be thought that the time has not yet been reached when a really worthwhile introduction to the New Zealand spiders could be prepared because only now are we beginning to know and understand our extraordinarily interesting, and in many ways, unique native fauna. But whatever the shortcomings of the present volume, it should at least serve to bridge the gap between the initial arousal of interest and curiosity and the interpretation of subsequent observations and so lead to a greater understanding. So little has been recorded that any enthusiast, whether professional or amateur, cannot help but discover new and previously unknown facts, and a study of these animals, which are both abundant and easy to keep in captivity, is well within the capabilities of biology students at secondary schools.

The absence of a store of published information to provide the basis for a book such as this has meant that many of the notes and observations are new but not all are our own. We have benefited greatly from the field observations and notes freely given by Mr C. L. Wilton and Professor B. J. Marples and are also deeply appreciative of the advice we have received from Mr C. L. Wilton during the preparation of the manuscript. To these and others who have helped in many ways we express our gratitude.

Our special thanks are due to Mr Barry Weston who has drawn most of the illustrations; we express our great appreciation for his patience and artistic skill in preparing illustrations which are both accurate and attractive.

Introduction

In the early days of the colonisation of New Zealand, certain spiders, such as the poisonous katipo, the marine spiders and the trapdoor spiders, were the subject of numerous notes published by the naturalists of that time but it was not until the last quarter of the nineteenth century that the abundance of these fascinating creatures became clear. A. T. Urquhart, who was farming at Karaka near Auckland published nineteen papers in the *Transactions of the Royal Society of New Zealand* between 1882 and 1897 in which he systematically described some hundreds of species. It might be thought that this pioneer study, which provided names for so many of our spiders, would have been used as a basis for further work on the natural history of the group. But, paradoxically, the sheer bulk of Urquhart's descriptions, coupled with the lack of sufficient clear illustrations, blunted the enthusiasm of all but the most persistent.

Twenty years passed after the last paper was published by Urquhart before another attempt was made to bring out a comprehensive survey of the New Zealand spiders. This was done by a Frenchman named de Dalmas. Comte Raymond de Dalmas visited New Zealand in 1912 in his sea-going yacht *Calazie* to indulge primarily in salmon fishing. He had, however, been persuaded by his friend Eugéne Simon of the Natural History Museum in Paris, at that time the world authority on spiders, to collect specimens for him. On his return to France, Dalmas was further persuaded and took over the study of specimens he had collected, and so in 1917 his *Araignées de Nouvelle Zélande* was published. This work, of over one hundred pages, gave a concise revision of the work which had been done before and recorded a number of new discoveries made by him. Despite the availability of this excellent introduction, a further twenty years passed before interest developed among resident naturalists. Fortunately, once this interest was aroused it persisted. Although the time is close when it should be possible to find the correct scientific name for most of the species which live in this country, at present barely one quarter of the two thousand or more species have been officially described. This lack of proper names for some of the species we mention raises a number of problems, particularly as we know that some of the names will be changed in the future. But here we are relying on illustrations to make clear which spiders we refer to.

If this book should arouse the interest of even a few to further the study of the lives and habits of our incomparable fauna, we shall feel that our efforts have been worthwhile.

PART I

CHAPTER ONE

Structure and Behaviour of Spiders

While much enjoyment and pleasure can be gained from the simple observation of spiders living their normal lives, it is useful to have some background knowledge of structure and typical behaviour patterns if an understanding of the purposes of these actions is to be achieved. Without this knowledge, behavioural characteristics, which might be readily recognized as such in larger animals like birds or mammals, are easily overlooked and much of the interest lost. Fortunately several of the books listed in the bibliography provide this information in some detail and as they are readily available only a brief summary is given below.

The body of a spider is clearly divided into two portions, the cephalothorax and the abdomen, which are connected by a narrow waist called the pedicel. The cephalothorax bears the chelicerae, palps, legs and eyes while the abdomen carries the respiratory, reproductive and digestive systems, as well as the spinnerets which provide the silk. The cephalothorax is composed of the head and thorax but these are always fused together as a single unit. The upper surface of the cephalothorax is covered by a single plate, the carapace, where a U-shaped groove may still show the limits of the head region which bears the eyes. Behind this groove, on the thoracic region, is usually a median depression called the fovea which marks the attachment of internal muscles. Most of the under surface of the cephalothorax is covered by a large plate, the sternum, which in turn has a small plate called the labium, hinged or fused in front. The eyes are simple ocelli and never exceed eight as this is the primitive number. They may be arranged in two or three rows near the front of the carapace and because their size and arrangement vary greatly, are used extensively to separate both species and larger groups of spiders. In some New

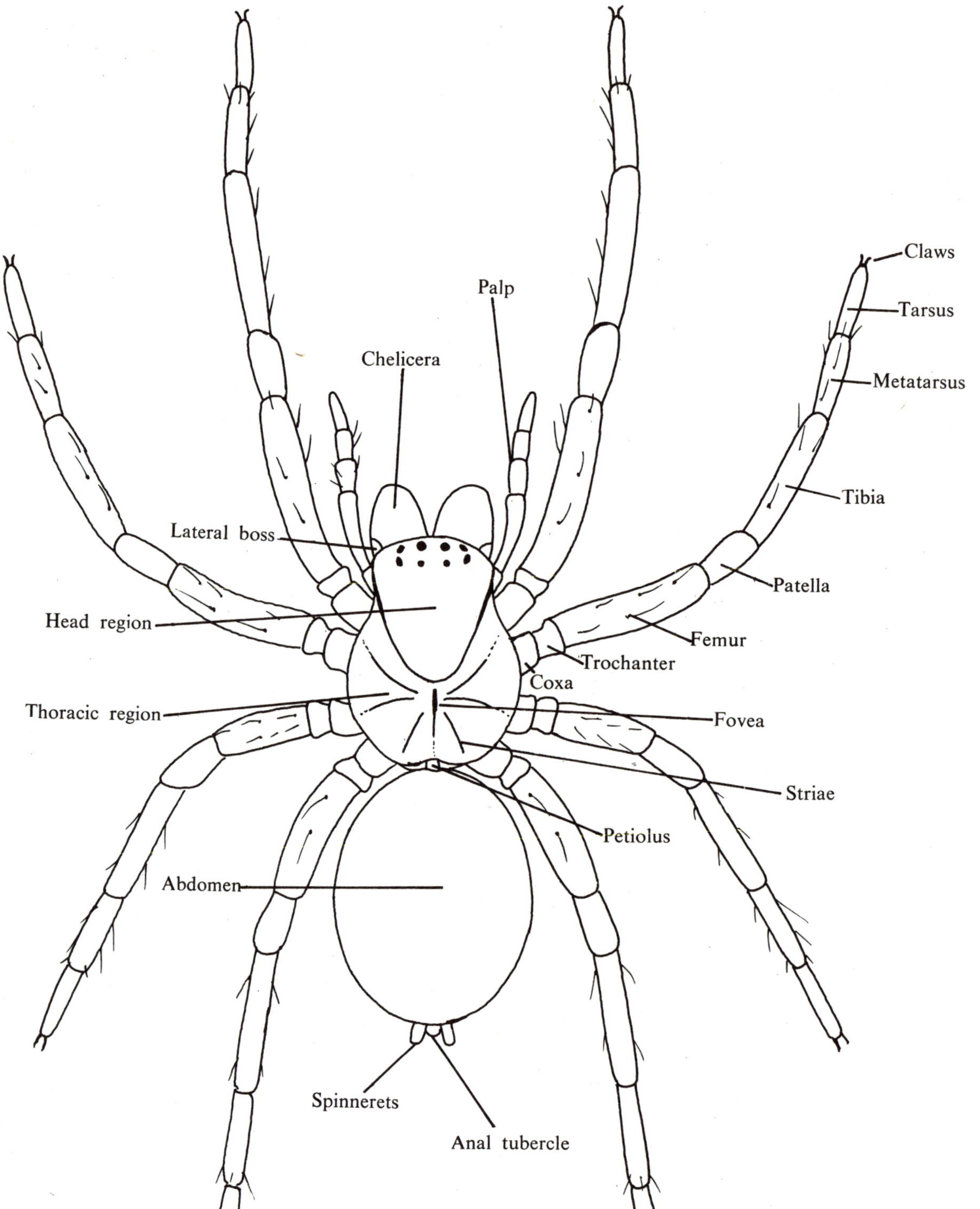

Fig. 1 Dorsal and ventral surfaces of a spider showing the main external features of a typical spider.

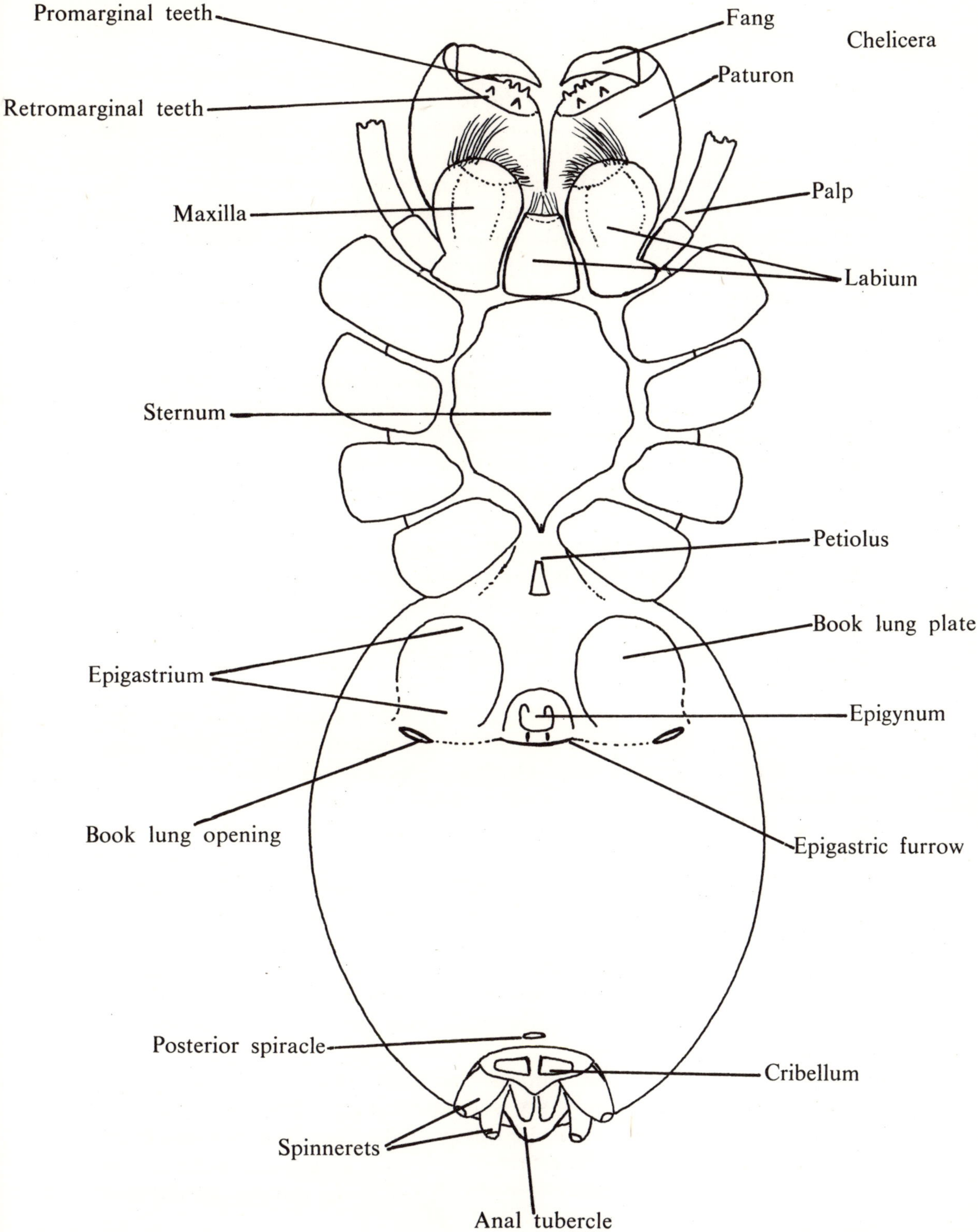
Promarginal teeth
Fang
Chelicera
Paturon
Retromarginal teeth
Palp
Maxilla
Labium
Sternum
Petiolus
Book lung plate
Epigastrium
Epigynum
Book lung opening
Epigastric furrow
Posterior spiracle
Cribellum
Spinnerets
Anal tubercle

Zealand spiders one pair of eyes may be missing and in all our six-eyed spiders this is the middle pair of the front row. Further reduction in the number of eyes is found in some overseas spiders where species with four or two eyes or even a single eye have been recorded. For lack of a better name, the area of the carapace between the front eyes and the margin is called the clypeus although this has no relationship to the structure more properly called by the same name in the insects.

The chelicerae are situated in front of the mouth below the front margin of the carapace and are easily seen when the spider is viewed from the front. Each chelicera consists of a stout basal segment and an apical fang which at rest folds down into a shallow groove on the basal segment. Most spiders use poison to subdue their prey and this is delivered by the chelicerae. In the mygalomorph spiders the poison gland fits within the basal segment but in other spiders it usually extends back into the cephalothorax, although it is completely absent in a few families such as the Pholcidae and the Uloboridae, which then must rely on silk to overcome their prey. The poison passes through a narrow duct within the fang which opens a short distance back from the tip. The poison glands are sheathed in muscle fibres so that the sac can be contracted as the spider bites and the poison thereby injected.

The mouth is a simple opening behind the chelicerae between the rostrum and the labium and leads to a strong muscular sucking stomach (Figure 2). Behind the sucking stomach the gut may be provided with numerous lobes in both the cephalothorax and the abdomen which probably act as food storage organs as well as producing some of the digestive enzymes which are exuded from the mouth while the spider is feeding. These enzymes break down the solid portions of the body of their prey into the liquid form it must assume before it is ingested.

The mechanical crushing of the prey to facilitate the action of the digestive enzymes is brought about by the fangs and the maxillae of the palps. The maxillae are plate-like extensions of the coxae of the pedipalps which extend forward to the side of the mouth and may be moved freely. A thick row of hairs on the inner margins of these plates, the scopulae, probably act as filters to prevent indigestible fragments from passing into the mouth.

The pedipalps are leg-like but as the metatarsus is always lacking these appendages have only six segments; coxa, trochanter, femur, patella, tibia, and tarsus. The pedipalp of the female is not modified in any particular way and the tarsus ends in a single small claw. The mature male, however, uses the pedipalps directly during mating and the tarsus which is often flattened or concave underneath carries a specialised structure, the palpal organ, which stores sperm derived from the gonopore on the ventral surface of the abdomen and eventually transfers it to the female during mating.

In its simplest form, the male palpal organ is merely a bulb with a distal projection from which a loosely coiled internal tube opens near the tip. Among the New Zealand fauna this type of palpal organ is found in the families Oonopidae and Segestriidae and all of the mygalomorphs (Figure 3). In more advanced spiders the palpal organ can become extremely complex and often accessary processes to assist in the orientation of the organ during mating are found on other segments of the palp.

Four pairs of legs are attached laterally to the cephalothorax between the sternum and the carapace. Each leg consists of seven segments from the base; coxa, trochanter, femur, patella, tibia, metatarsus, tarsus. The trochanter is usually the shortest segment and the femur longest. In some families, such as the Thomisidae, the front two pairs of legs are twisted back so that they are directed to the side and the normal front side of the legs becomes the dorsal surface. When this is the case, the spiders are called laterigrade. All legs are clothed with hairs and many have stronger spines and bristles. The claws are used for clinging to the silk of snares and most web building spiders have three – two upper claws and a single lower claw. In many of the hunting spiders only the upper pair is present and then a dense clothing of short hairs, the scopula, is often found on the ventral surface of the tarsus and sometimes on other segments. The scopulae may assist the spider to gain purchase on the surface as it moves and it is the effect of these scopulae spreading the weight which enables the water spider to skim over stretches of water without breaking through the surface film.

Many sensory structures found on the legs provide the receptive organs which stimulate certain aspects of their behaviour. Long and slender erect hairs which probably detect vibration are present on the legs of most spiders, usually on the tarsus, metatarsus and tibia. These trichobothria are each set in a large socket on the upper surface of the segment and usually are easily seen. A small blister, opening to the exterior through a small apical hole, is found on the tarsi of many spiders. The function of this structure, called the tarsal organ, still remains a puzzle although it has been suggested that as a chemoreceptor it plays a particular part in the detection of water. It is, however, not present in all spiders and so its function must be shared with other organs. Groups of minute slits provided with nerve endings can be

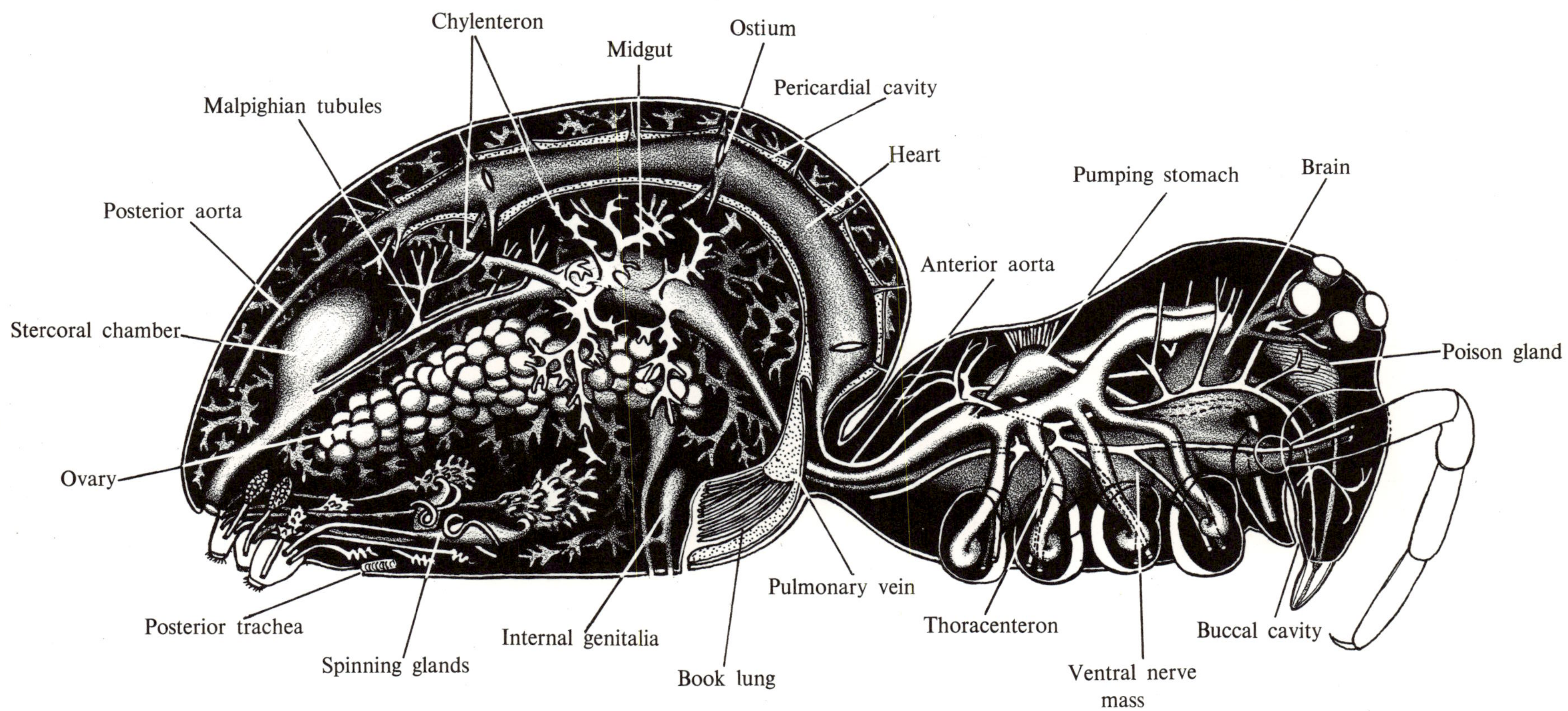

Fig. 2 The main internal organs of a female spider.

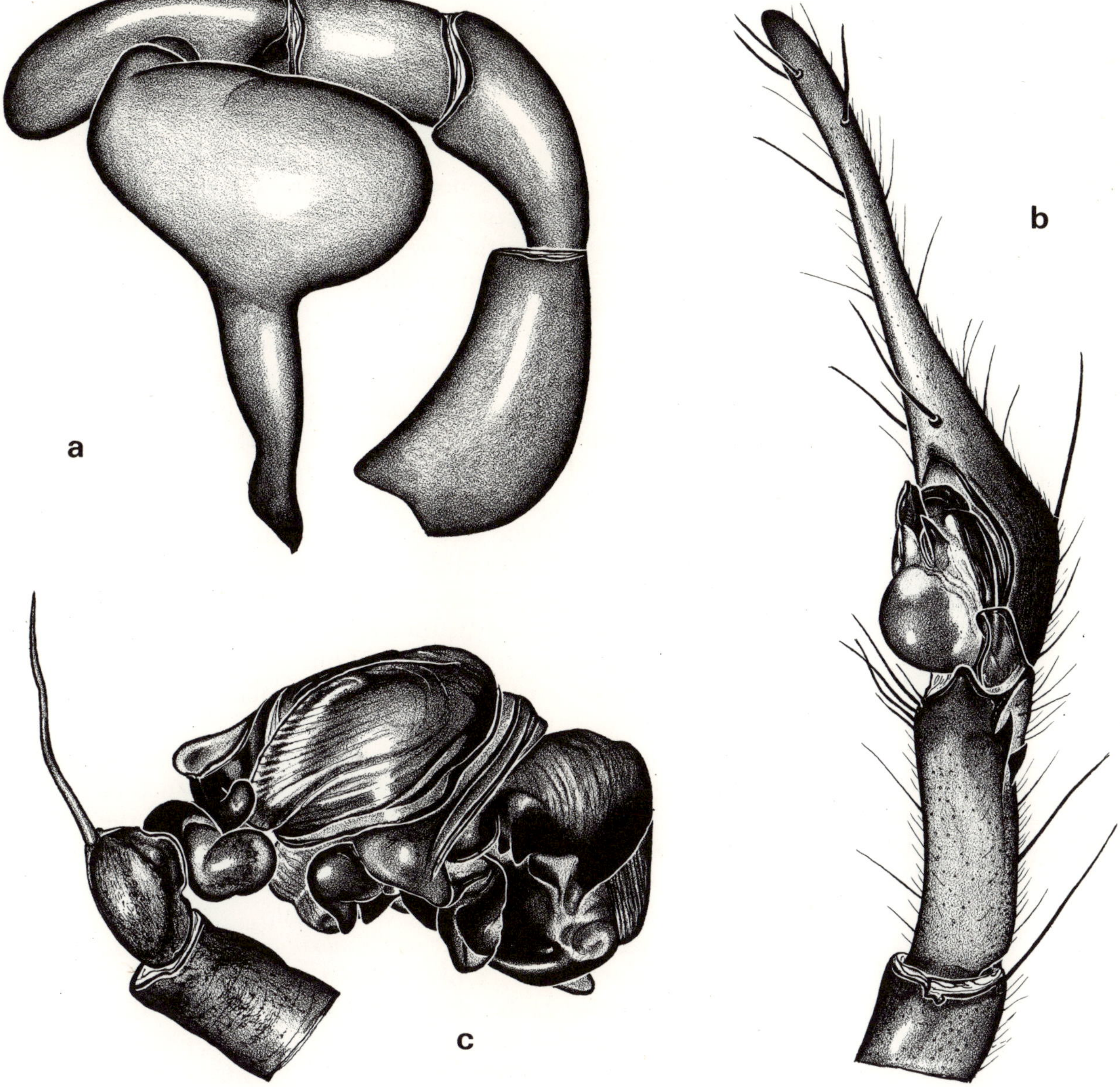

Fig. 3 The male palpal organs of three families showing the increasing complexity of the structures. a. *Duripelta* sp. Family Oonopidae. b. *Cambridgea antipodiana*. Family Agelenidae. c. *Aranea pustulosa*. Family Araneidae.

seen in the integument of the legs and to a lesser extent on other parts of the body. Experiments in the past have suggested that these structures, the lyriform organs, might be chemoreceptors but more recently it has been concluded that they help in detecting torsion and stress, and, by relaying the relevant information to the musculature, play an essential part in regulating the animal's posture. There are numerous other structures found on the legs of restricted groups of spiders which undoubtedly have some function but little investigation of their uses has been made. Hairs and bristles are often modified for use in special circumstances when their regular arrangement sets them apart from the normal random covering. In addition to the cribellum described later as a spinning organ, the cribellate spiders have one or two rows of closely-spaced curved bristles along the metatarsus of the fourth pair of legs. This is called the calamistrum which is used by these spiders to comb out on to silk the special sticky fluid exuded from the pores of the cribellum. Most of theridiid spiders have a row of strong serrate bristles or comb on the ventral surface of the tarsus of the fourth pair of legs. This tarsal comb takes up the swathing silk from the spinnerets flung by the fourth pair of legs over prey trapped in the web. In a number of different groups of spiders a distinctive row of stout bristles is arranged across the distal end of the metatarsus of one or more pairs of legs. The function of these structures, which are often called preening combs, is unknown.

Stridulating structures which are found in a number of male spiders function on a scraper and file system. In some theridiid spiders, such as many species of *Steatoda*, the file is situated on the hind portion of the cephalothorax and the scraper on the opposing portion of the abdomen, while in many linyphiids the file is found on the outer surface of the chelicerae and the scraper on the inner surface of the palp.

A most unusual stridulatory system is found in *Cambridgea antipodiana* where the scraper is on the dorsal surface of the waist of the spider and the file is a concave, ridged plate on the abdomen (Figure 4). The thin waist which joins the cephalothorax to the abdomen is called the pedicel. The abdomen is usually oval but may be modified in many ways. Sometimes it is elongate but this may have developed in two different ways. In some spiders such as *Tetragnatha* the whole abdomen is elongated so that the spinnerets are still at the end, but in other spiders, such as *Arachnura*, the abdomen may have become elongated above and behind the spinnerets so that these are now on the under surface and then the posterior portion may be called a tail (Figure 101). The dorsal surface may also be raised sharply so that from the side the abdomen

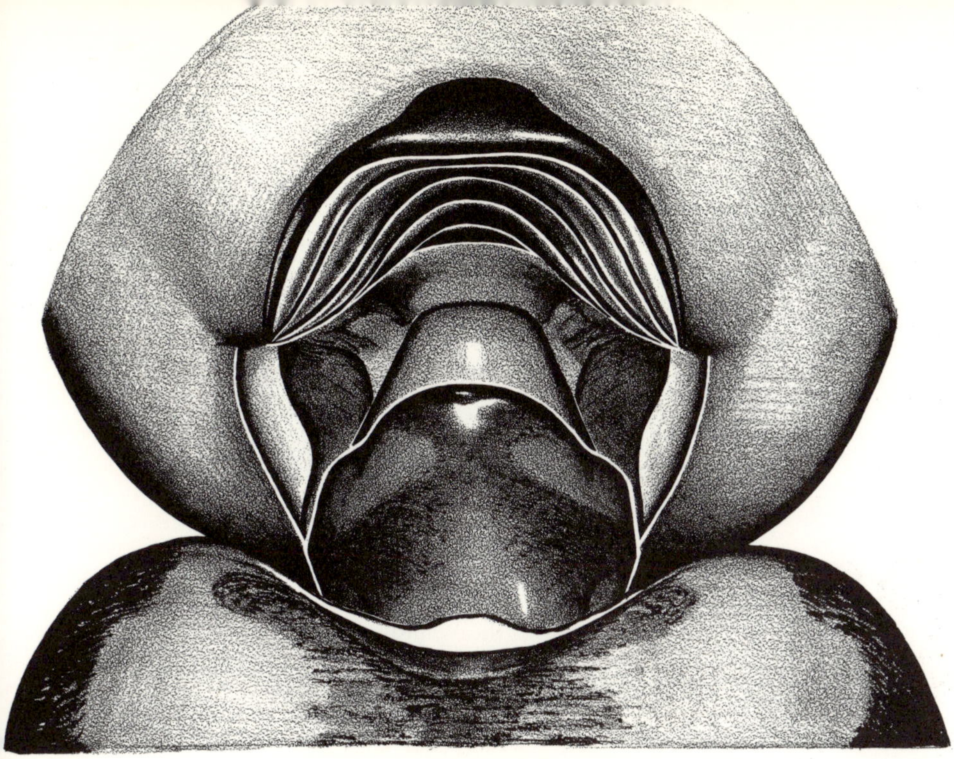

Fig. 4 Stridulating organs. a) Male of *Cambridgea antipodiana*. A spur on the waist rubs on the ridges of a plate on the abdomen. b) Male of *Steatoda* sp. A hard projecting plate on the abdomen rubs against ridges on the carapace.

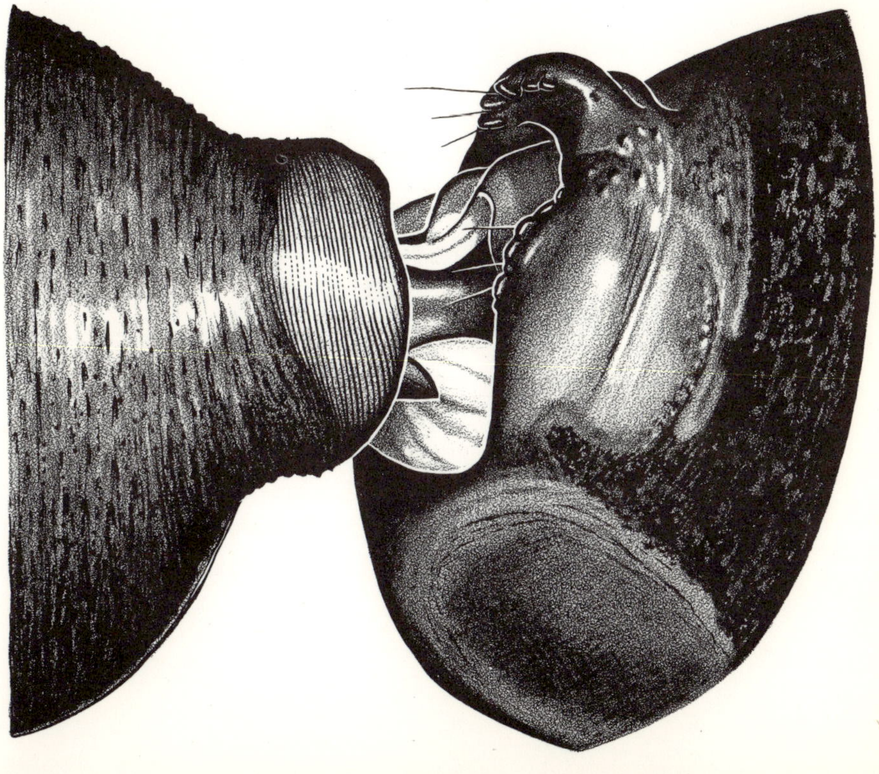

is triangular as is characteristic of most species of *Tekella* and *Rhomphaea*, or the abdomen may be ornamented with protuberances on the upper surface as in *Celaenia* and *Phoroncidia*. In many of the midget spiders of the family Symphytognathidae the abdomen is protected by hard plates or scutes which for some reason are normally more extensive in the males.

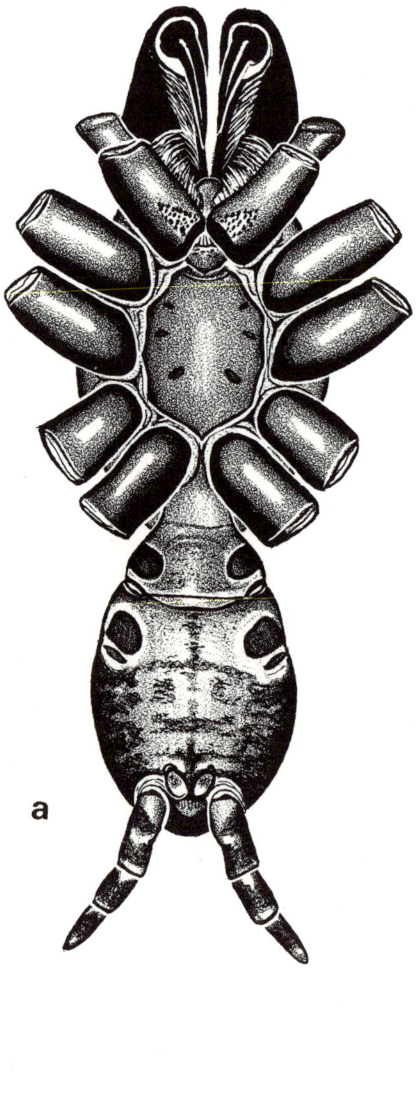

a

Fig. 5 The ventral surface of four spiders showing the appearance of paraxial and diaxial chelicerae and the different forms of the respiratory organs. a) *Porrhothele antipodiana*. Family Dipluridae, with paraxial chelicerae and four book lungs (shown with dark shading). b) *Gradungula sorenseni*. Family Gradungulidae, with diaxial chelicerae and four book lungs. c) *Clubiona* sp. Family Clubionidae, with diaxial chelicerae, one pair of book lungs and a single spiracle near the spinnerets leading to tracheae. d) *Dysdera crocata* Family Dysderidae with diaxial chelicerae, one pair of book lungs but with a pair of spiracles immediately behind the epigastric groove leading to tracheae.

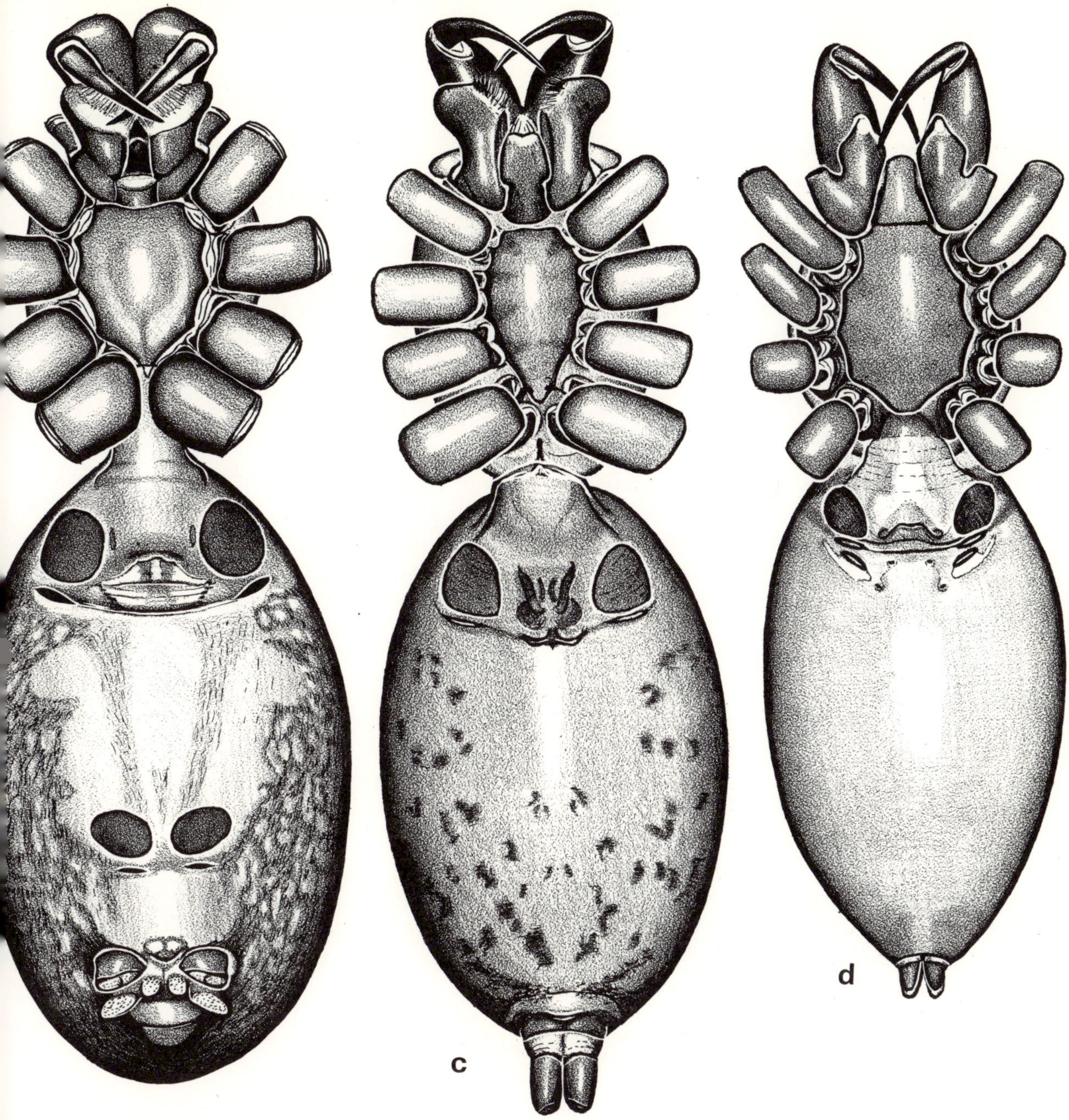

The ventral surface of the abdomen is divided by a transverse groove called the epigastric furrow which marks the division between the second and third abdominal segments. The second segment extends from the petiolus to this groove while the third segment extends back from the groove to the spinnerets so that the structures associated with these segments are always found within these limits. Associated with the second segment are the anterior pair of book lungs and the reproductive system, while the second pair of book lungs or the posterior tracheae open from the third segment. There are two groups of four lunged spiders found in New Zealand; the mygalomorphs of the families Ctenizidae, Dipluridae and Migidae and the peculiar true spiders of the family Gradungulidae. By far the greatest majority of spiders have retained the anterior pair of book lungs but the posterior pair have been replaced with tracheae (Figure 5). In some spiders such as the Oonopidae and Segestriidae a pair of slits leading to tracheae is found immediately behind the epigastric groove but more often these paired tracheae have fused so that there is a single median spiracle which may be placed anywhere between the spinnerets and the epigastric groove but more commonly is situated immediately in front of the spinnerets.

There are a few spiders, represented in New Zealand by the Symphytognathidae, where the anterior pair of book lungs has been replaced by tracheae and so these spiders have no book lungs at all. The book lungs are always in the form of a blind pocket, the floor of which is folded into a number of overlapping leaves, giving the organ its common name. Pulsations of the heart move the exhausted blood freely through the inner cavity of these leaves enabling an exchange of oxygen through the outer surface where small projections keep them slightly apart so that air can be circulated within the vestibule of the lung. The tracheae divide and extend from the spiracles to deliver oxygen directly to the organ they serve. In general, the respiratory system in spiders evolved through a gradual change from book lungs to tracheae. Although in the past doubts have been raised about the possibility of such a direct evolutionary change from one form of repiratory system to another, recent investigations have shown that the book lungs of some living spiders have a reduced number of greatly elongated leaves which could represent an intermediate stage in the transformation from book lungs to tracheae. It seems probable that the evolutionary trend within spiders as a group for the book lungs to be transformed into tracheae was concerned as much with the need to conserve moisture within the body as with more efficient oxygen intake. The book lung system presents a very large surface directly to the air through which water may be lost. The ventral surface of the abdomen in front of the epigastric groove is not modified in the male but in the female it has a distinctive plate called the epigynum (Figure 6) in many of the more specialized families. The development of the epigynum is related to the increasing complexity of the palpal organ of the male and so we get an association in such spiders as the Oonopids and Mygalomorphs of a simple palpal organ and no epigynum. If, as seems probable, the tendency has first been for the male palpal organ to become more complex (Figure 3), then the development

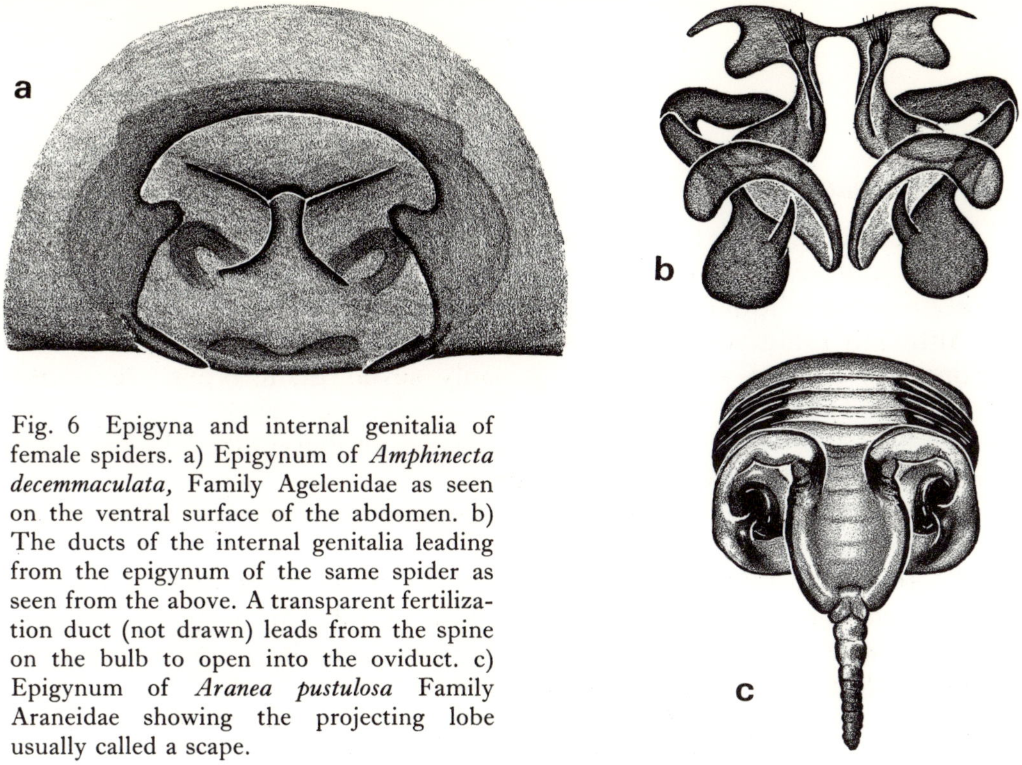

a b c

Fig. 6 Epigyna and internal genitalia of female spiders. a) Epigynum of *Amphinecta decemmaculata,* Family Agelenidae as seen on the ventral surface of the abdomen. b) The ducts of the internal genitalia leading from the epigynum of the same spider as seen from the above. A transparent fertilization duct (not drawn) leads from the spine on the bulb to open into the oviduct. c) Epigynum of *Aranea pustulosa* Family Araneidae showing the projecting lobe usually called a scape.

and increased complexity of the epigynum can be looked upon as a complementary device to ensure that the tip of the embolus through which the sperm is passed into the female will reach the correct place during mating, although surprisingly the epigyna of a number of spiders with complex palps may be quite simple. The basic function of the epigynum is to provide access through a pair of apertures to an internal receptacle which stores the sperm until such time as it is required to fertilize the eggs passing down the oviduct. Narrow fertilization ducts to carry the sperm at this time connect the receptacula with the oviduct. The epigyna are often ornamented with ridges, grooves and spines which engage and guide various parts of the palpal organ and processes on the segments of the palps during mating and so there is a fairly close correlation between the structure of the two organs. Even spiders with no epigynum have internal structures associated with the oviduct but these do not necessarily receive the sperm which, in these spiders, may be stored in the oviduct where it is directly deposited during mating. At the end of the abdomen, a lobe called the anal tubercle marks the exit of the alimentary canal; it is below this structure that the spinnerets are found. While it is thought that eight spinnerets is the original number, only a few rare species still possess this number and the normal number on recent spiders is six. Some New Zealand spiders have only one or two pairs of spinnerets but the majority have three pairs which are arranged so that the middle pair, which is small, is almost hidden by the other two pairs. When the spinnerets are examined under magnification numerous hollow rods, the spigots, can be seen. It is from these minute structures that the silk is exuded from glands within the abdomen as a liquid which rapidly hardens into a thread as it is produced. Although hundreds of glands may be present in a single spider, each opening from a separate spigot, only seven main types are found. (Figure 7).

1. *Aciniform Glands* Usually associated in berry-like clumps, each gland is spherical with a short duct.
2. *Pyriform Glands* They are also found in clumps but longer and narrower than the Aciniform glands.
3. *Ampullate Glands* These glands are never numerous and are in the form of a cylindrical tube of which the middle part is greatly enlarged.
4. *Tubular Glands* These are cylindrical glands, more or less even in width and sometimes convoluted.
5. *Aggregate Glands* Irregularly branched and lobed glands which bear cellular knobs on the duct.
6. *Lobed Glands* These glands which are found only in the Theridiidae consist of a mass of irregular lobes and do not have knobs on the duct, the characteristic of aggregate glands.
7. *Cribellum Glands* Small spherical glands which open through the pores of the cribellum.

The aciniform glands which open from the median and posterior spinnerets produce the swathing band used by numerous spiders and possibly provide silk for the egg-sac of some spiders. The pyriform glands are associated with the anterior pair of spinnerets and produce the silk used in the attachment disc. The ampullate glands are never very numerous, ranging from two to six pairs which open from the anterior and median spinnerets. These glands seem to be present in all spiders and provide the silk for the drag-line and the dry threads of webs. The cylindrical glands are commonly found in female spiders but are reduced or absent in the males. The six glands usually present in females open from the median and posterior spinnerets and are used to produce the silk for covering the egg-sacs. The aggregate glands are found only in the Araneidae, Theridiidae and Linyphiidae where six of these glands open from the posterior pair of spinnerets. It is thought that these aggregate glands produce the sticky threads used in their snares. The lobed glands are found only in the Theridiidae where they are used to produce the special swathing silk which these spiders fling over their prey with the tarsal comb of the fourth pair of legs. The cribellum glands secrete the silk which is combed out with the calamistrum (Figure 8) on the metatarsus of the fourth leg of cribellate spiders to produce the

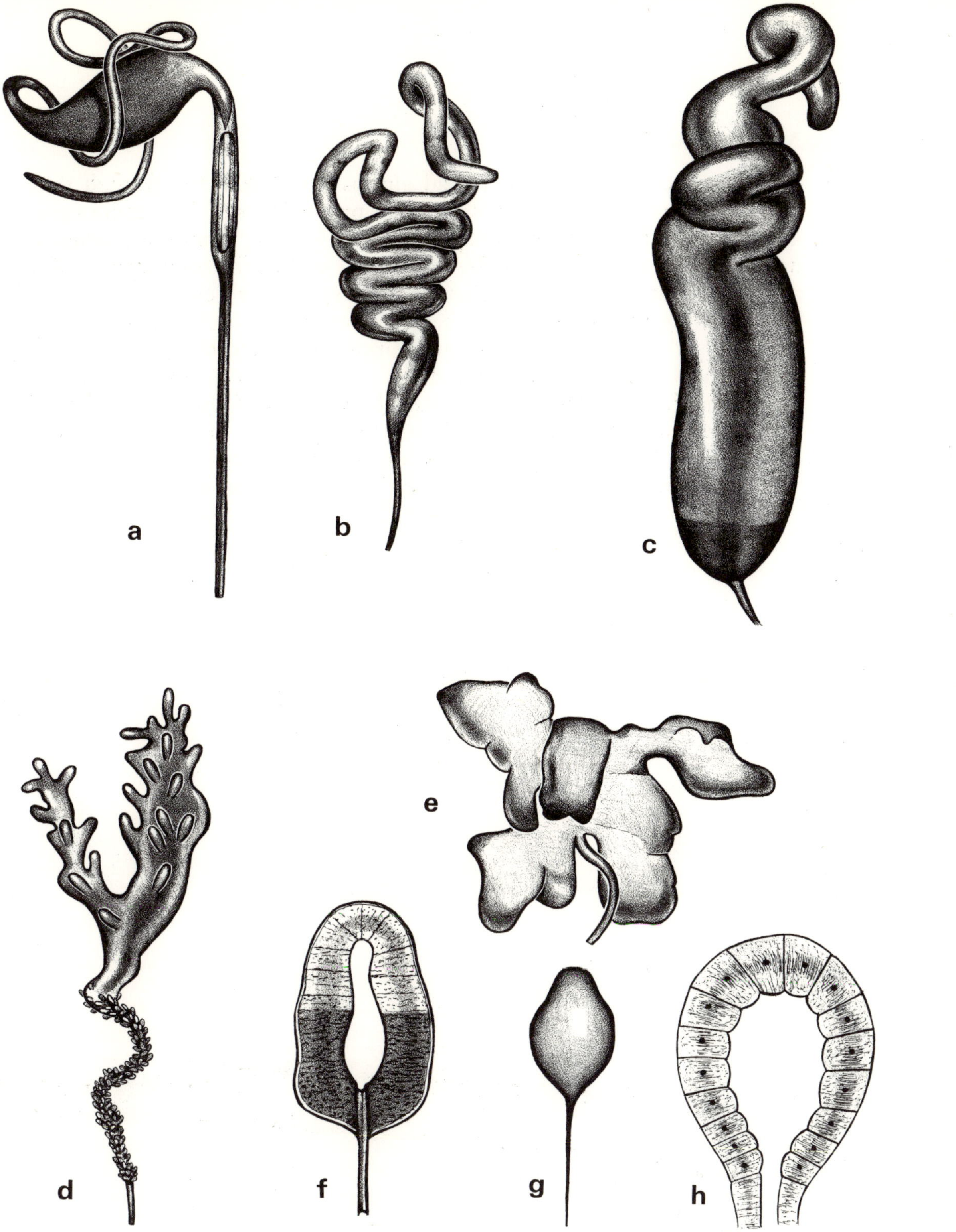

Fig. 7 Silk glands. a) Ampullate gland. b) Tubular gland. c) Ampullate gland. d) Aggregate gland. e) Lobed Gland. f) Section of a Pyriform gland. g) Aciniform gland. h) Section of an Aciniform gland.

characteristic hackled band. It had always been taken for granted that the spinnerets were the only structures producing silk in spiders until in 1964 Dr M. Melchers published in a German scientific journal a description of glands which she had found within the epigastrium of a male mygalomorph spider opening from spigots which had probably been seen previously but had been considered to be ordinary hairs. She found that these glands produced the silk which the spider used to spin a small sheet on which the drop of sperm was deposited while it was recharging its palp. Professor B.J. Marples of the University of Otago found these glands were extremely well developed in the *Gradungula* and his further study demonstrated their presence in a wide range of families including many of the true spiders.

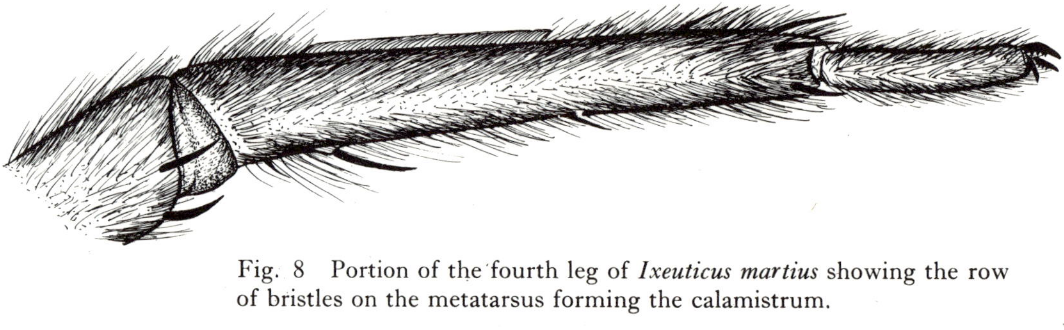

Fig. 8 Portion of the fourth leg of *Ixeuticus martius* showing the row of bristles on the metatarsus forming the calamistrum.

CHAPTER TWO

The Life of a Spider

While not wanting to get involved in the usual argument about which came first – the hen or the egg – we feel like most chroniclers of life histories that it is most convenient to start with the egg. Although a few small spiders, such as the rarely seen six-eyed Oonopids and the minute Symphytognathids, are able to limit the numbers of eggs they lay each time to perhaps only two or three, the hazards of life for the majority of spiders are usually so great that each female must produce hundreds or even thousands of eggs each season to ensure that their line persists through to the coming year (Figure 9). Some of these spiders such as the common daddy-long-legs (*Pholcus phalangioides*) bind the eggs together with little more silk than is needed to keep them from spilling out but more commonly they are securely enclosed and protected by one or more layers of special silk to form a characteristic egg-sac. The egg-sacs themselves are sometimes similar within a whole family as in the Wolf spiders and the Nurseryweb spiders while in others a wide range is found, but the egg-sacs produced by each species vary little one from another. Very few of the egg-sacs of New Zealand spiders are known and many unusual constructions which have been found still await identification because frequently the spider abandons them after they are constructed, leaving behind no clue to ownership.

In general, it seems to take from three to four weeks for the spiderlings to develop within the egg, although the climatic conditions cause variation and there are those spiders which seem to require a longer time. None has been recorded overwintering on the egg stage. Once the spiderling has completed its development within the egg, it must still break out from the thin, but strong, membrane in which it is enclosed and to assist with this, the first of its continual struggles for survival, it has a small transitory spine on the base of each palp. The newly hatched spiderling is completely helpless; it is unable to feed or even to

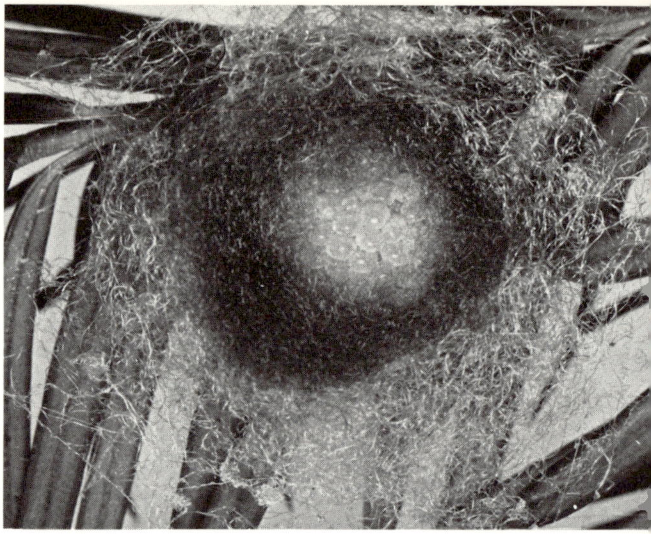

Fig. 9 *Aranea pustulosa* a) The mother guards the egg-sac. b) Newly laid eggs. c) After three weeks the spiderlings hatch and shortly pass through the first moult.

spin silk and is dependent at this time for nourishment on the remnant of the egg yolk enclosed within its body during development, and so for some time it will remain secure within the silken cover which earlier protected the eggs. At least one, but more usually two, moults will need to be passed through before the spiderlings are sufficiently developed to lead an independent life so that at least a week and often considerably longer will pass before they emerge. There may also be a time after they have broken from the egg-sac when they are still protected by the mother in a number of ways. Young wolf spiders clamber from the egg-sac to the mother's body where they cling for some days while the nurseryweb spiders carefully construct a silken web around the egg-sac which shelters the spiderlings after they emerge. The young of many of the tube and tunnel dwellers, such as the trapdoor spiders and the six-eyed *Ariadna*, may remain in the burrow with the female for some considerable time during which it is presumed they are permitted and perhaps assisted to share the food caught by the mother. A truly advanced form of parental care has been recorded for one of the European cobweb spiders *Theridion sisyphium* where the female actually feeds her young and then later shares with them the food she catches. Bristowe (1958) describes the behaviour of this spider in the following words;

> On a sunny day soon after emerging from the tent in which they hatched from their egg-sac the mother can be seen hanging downwards with the babies jostling each other to get near her mouth from which a drop of fluid is oozing. Here two or three of them at a time will have their first meal from their mother's mouth . . . Mouth feeding continues for several days and then the young begin sharing her meals. I have seen the mother and perhaps thirty babies all feeding at once on a large fly. That this is not just an apathetic failure on her part to drive them away is shown by her departure from normal practice when she has captured the fly. Usually she bites the insect in one or two places and sucks its body fluids through these punctures, but when she has young whose chelicerae might be too feeble to pierce an insect's chitin she passes over its body biting it freely in many places.

We have looked for the same behaviour among the scores of similar spiders in New Zealand as yet in vain but there is no reason not to believe that some of our own species behave in the same way. Once the spiderlings have passed through their early moults and can live their own lives they disperse rapidly. The sedentary spiders seem to wander off and often settle close to where they were born, but many of the web spiders take advantage of the wind to carry them far and wide. When these spiders "come of age" and are ready to leave home they clamber to the tip of a twig exposed to the breezes. There they stand on tiptoe as they let out a tuft of silk until the pull of the wind is strong enough to waft them through the air (Figure 10). The term ballooning is usually used to describe this habit; the threads they use form the gossamer which in some countries in certain seasons comes to rest like a light fall of snow. This is a scene uncommon in New Zealand because only a small proportion of our spiders has this habit. Possibly ballooning is not very effective in the thick forest where the bulk of our spiders live and in such a narrow ocean-bound land would result in tremendous losses with those blown out to sea.

Like all arthropods, the spiders are enclosed in a firm outer skeleton and so apart from the elasticity of the abdomen they are not able to grow gradually bigger as they get older. They must shed the hard outer

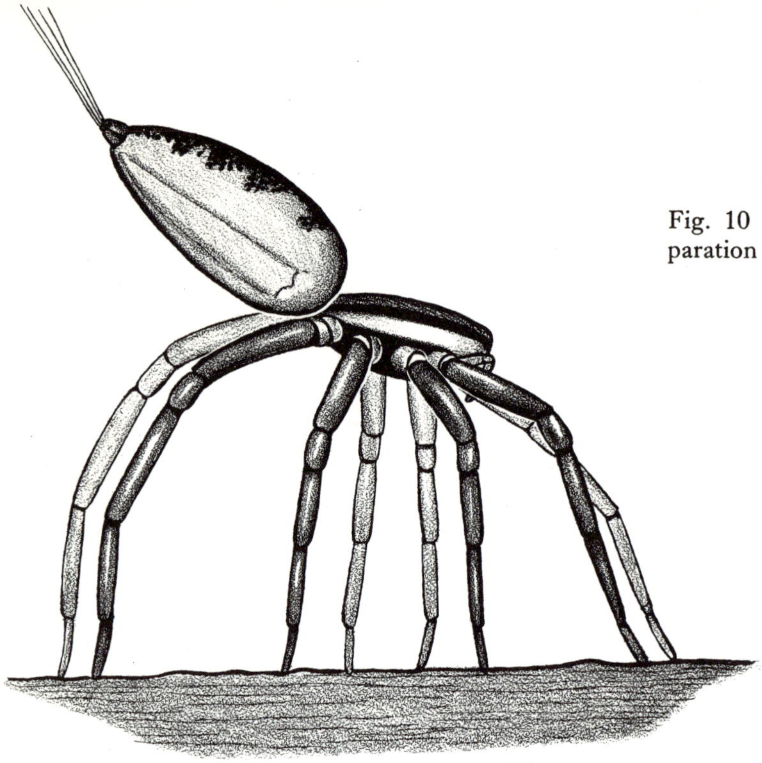

Fig. 10 Spiderling letting out silk in preparation for ballooning.

skin at intervals and then expand rapidly before the new exposed skin has hardened. This process of shedding the skin is known as the moult or ecdysis and the form the spider assumes between ecdyses is called an instar. It is characteristic of spiders that there is never any great change in form between instars apart from the sudden increases in size of the cephalothorax and appendages. The method of moulting is similar in all spiders (Figure 11). A few days before the moult takes place, the spider ceases to feed and becomes sluggish. At this time, some of the tube dwellers seal up the opening of their home while many of the web dwellers suspend themselves upside down from their web with the legs grouped together, and still others may lie on their back or side. Apart from this type of variation, the methods used by the spider to rid itself of the unwanted skin are remarkably stereotyped. First the skin below the carapace above the legs splits and the carapace lifts off like a lid but remains attached to the petiolus. The skin of the abdomen then splits along the sides and as the abdomen comes free the legs, palps and chelicerae are drawn out with a series of rythmical movements like fingers from a glove so that eventually the spider is hanging from the tip of its abdomen. At this stage the newly moulted spider is soft and defenceless and thus it must remain until the skin has hardened. However it is not entirely inactive during this time because the appendages must be repeatedly flexed or they will become stiff and distorted, and the spider will not be able to carry on its normal life.

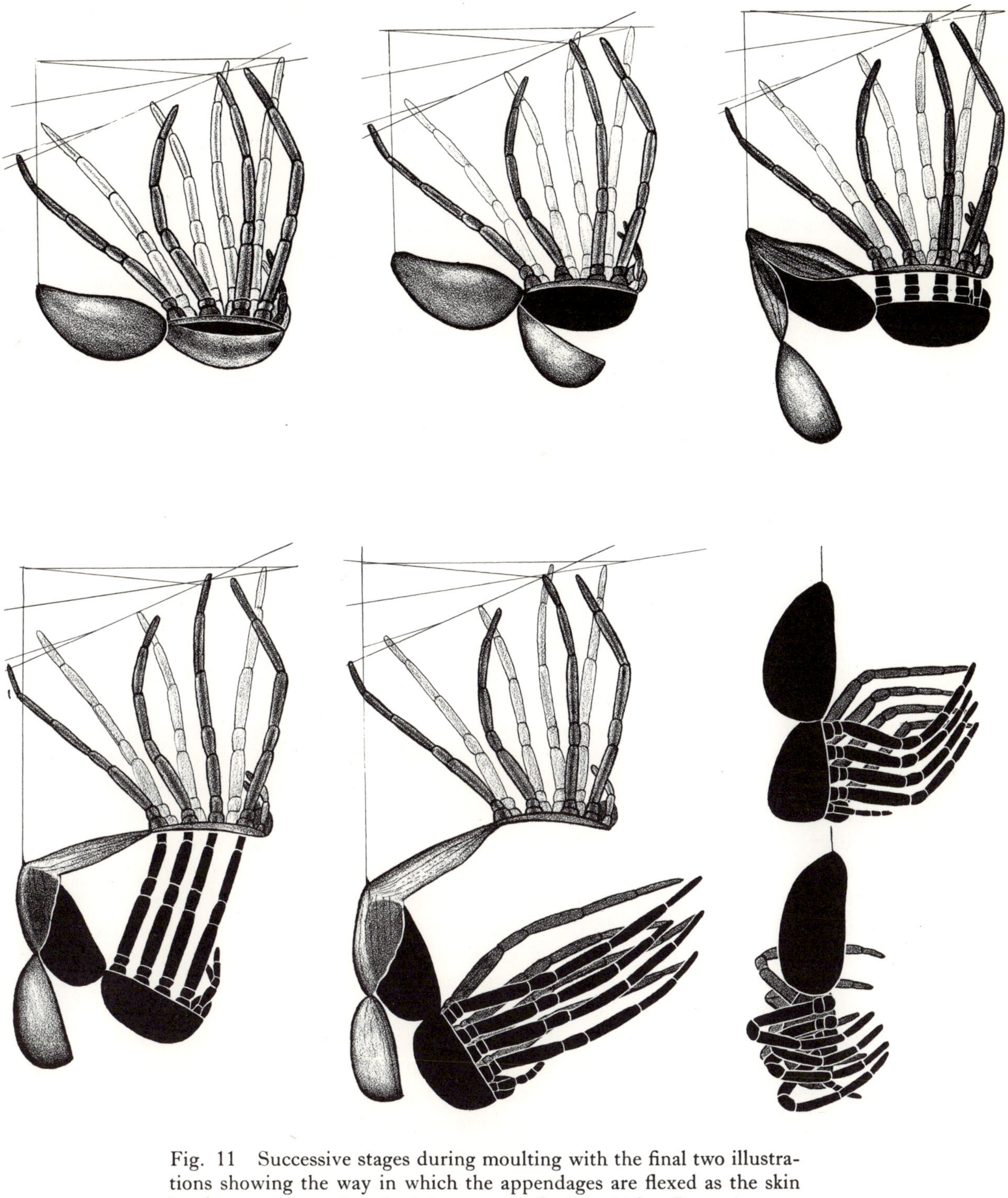

Fig. 11 Successive stages during moulting with the final two illustrations showing the way in which the appendages are flexed as the skin hardens to ensure that the joints remain flexible. (After Bonnet)

The number of instars a spider must pass through to reach maturity seems to vary with size, not only between species but also between the male and female of the same species in which the male is much smaller than his mate. Often males would appear to take advantage of their early maturity, since they may be found waiting with the female inside her retreat as she undergoes her final moult and are then able to complete mating while she is still helpless. Among our own spiders this is commonly seen with species of *Clubiona* and some of the Gnaphosids and surprisingly it seems to be common practice with the large jumping spider *Trite planiceps*.

It is only after the final moult that the secondary sexual characters are fully in evidence. The epigynum of the female and the male palpal organ are then present while the full colour and ornamentation which are often emphasised in the male are revealed.

During all stages of their life-history, appendages may be lost but unlike those of the harvestmen these are subsequently replaced. The joint between the coxa and trochanter is always weak so that even a slight pull on the leg may result in it dropping off but the break is then quickly sealed by a film formed from the blood, so that little body fluid is lost. If the appendage is ruptured elsewhere than at this joint, the spider will pull the remains of the leg off by tugging at it or by first fastening it with silk so that it may be pulled off. Undoubtedly the ability to lose a leg without any permanent inconvenience is an excellent adaptation for survival when attacked by a predator but this is by no means the complete story because all the appendages including the chelicerae and spinnerets are capable of regeneration which is a feature common to most arthropods. Much of our knowledge of regeneration in spiders comes from the extensive experiments by the prominent French araneologist Pierre Bonnet, who even went so far as to demonstrate that after he had removed all the limbs of a spider it could still regenerate them, although this would be of little use to the spider in nature because Professor Bonnet had to feed it carefully by hand as it rested in a soft bed of cotton-wool. The new limb grows coiled within the coxal segment until the next ecdysis when it extends as the old skin is shed. However, the limb must be lost during the first quarter of an instar if development is to take place and if it does develop it will not be as long as the normal segments. At least three moults are required before the appendage reaches its full size. The regeneration of the female pedipalp presents no difficulties because in the adult form it is not modified but the male palp with its complex mating organ requires three moults to finish up fully formed and functional.

Silk is the substance which ramifies throughout the whole of the spider's life and no other group of animals has exploited the possibilities of this versatile material to anywhere near the same degree. During its very first excursion, the spiderling lays down the dragline on which it will dangle if it loses its footing; by this line it will drop from twig to twig as it moves along. Perhaps this habit of relying on a safety line lies at the root of the evolution of the multitude of ingenious snares which these spiders today use to catch their prey. A superficial survey of the forms of snares used by spiders now can be used to show how many of these seemingly divergent forms of web may have developed. It is probable that the dragline has two primary functions, first as a safety line and then as a guideline to lead the spider back to its lair. As the spider moves along, the dragline is attached at intervals so that there is never a great length of the thread free. From this general habit, it is a very small step to the six-eyed *Segestria* living in tunnels in twigs, where a series of radiating threads act as triplines to warn the spider when prey is near and also to indicate the direction in which it lies. A further elaboration of these threads would give the wide-mouthed extension constructed by the tunnelweb spider *Porrhothele* or its counterpart in the true spiders with a sheetweb leading to a tube retreat, as in the cribellate black house spider *Ixeuticus martius* or the European house spider *Tegenaria domestica*. With the evolution of flying insects, the spiders would have been able to migrate very profitably above the ground into the plant cover and it is the aerial webs which demonstrate the vast range of possible elaborations. The simplest kind of snare is a single thread; such a web is constructed by the minute cobweb spiders of the genus *Phoroncidia* which are common in all our forests. Most of the theridiid spiders construct a seemingly higgledy-piggledy maze of threads, some of which may be sticky. Many of the Linyphiids, *Cambridgea* and others construct a sheet below the maze which then acts as a knockdown area for flying insects while the spider waits below the sheet to pounce on the insects as they fall. The peak of efficiency among the aerial spiders is attained by the cobweb spiders whose cartwheel webs combine strength with simplicity, but there is no indication among living spiders of the path which was traced to reach this form.

Even when the prey is entangled further, steps may have to be taken before a meal is assured; hence we find that many spiders have developed ingenious methods to ensure that the prey is not lost. Theridiid spiders fling sticky swathing silk with the special comb at the end of their fourth legs while the cobweb spiders wrap the victim in a thick sheet of silk (Figure 12). Many spiders vibrate the whole web so that the prey becomes more firmly enmeshed.

The path towards the perfection of snares does not always lead to bigger and better ones. This is shown by some of the cobweb spiders which have forsaken the formal snares for other methods. The two species of Australian orbweb spiders in the genus Dichrostichus do not construct any snare at all but instead let down a short thread of silk from the second pair of legs. When a moth approaches, the thread is twirled around until the moth is caught by a sticky drop at the end. These spiders are also represented in America and South Africa but do not occur in New Zealand. However, the more distantly related *Celaenia* which achieves the same results by simply grasping the moths with its front legs as they come close, is fairly common. The abandonment of the certainty of the orbweb for this seemingly precarious way of life may appear to be rather unusual but as is so often the case, a further factor must be considered, because these spiders may actually attract male moths to their doom by a scent or odour. Many hunting spiders use silk in some way to capture their prey. The rapidly moving gnaphosids use a strong band of silk which they stream out as they attack and so pin their victim down as they close in for the kill while the extraordinary

Fig. 12 *Aranea pustulosa* wraps up its prey.

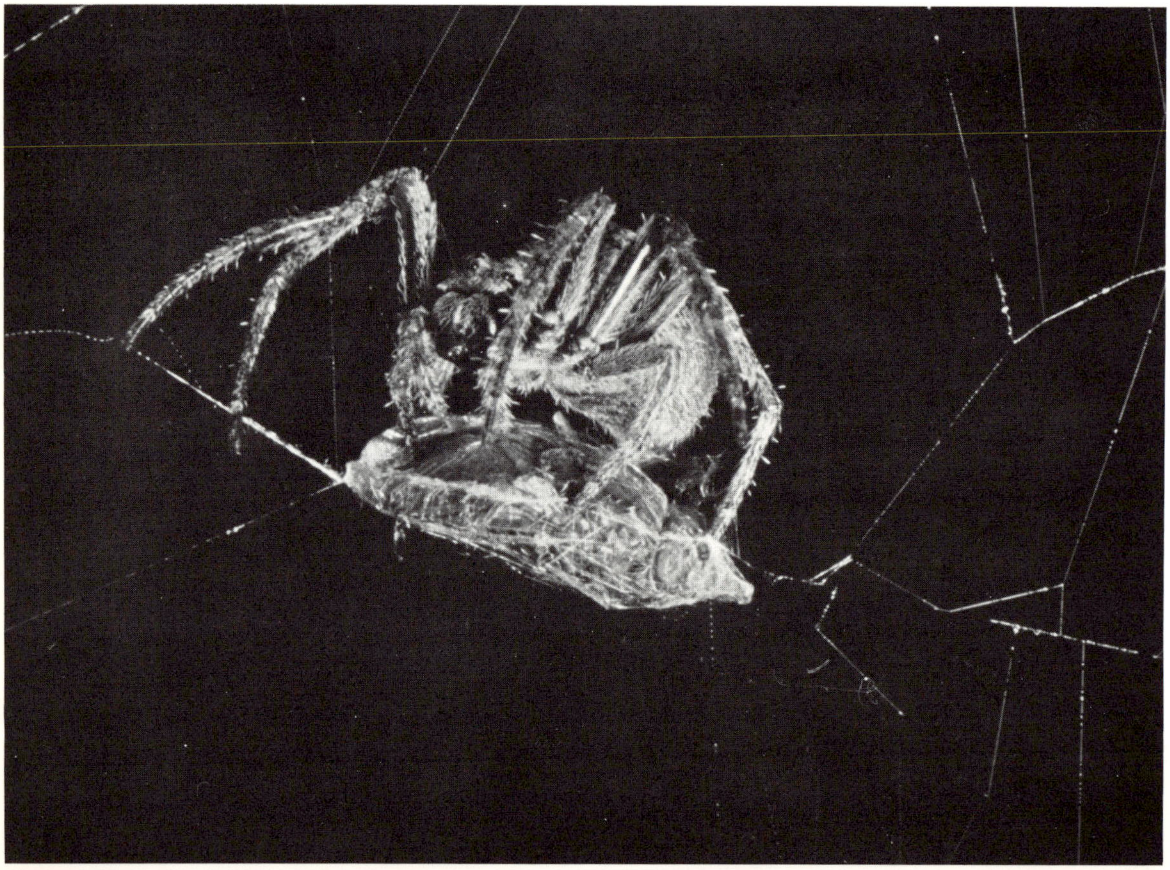

Fig. 13 Male spider charging the palpal organ from a sperm web (after Bonnet).

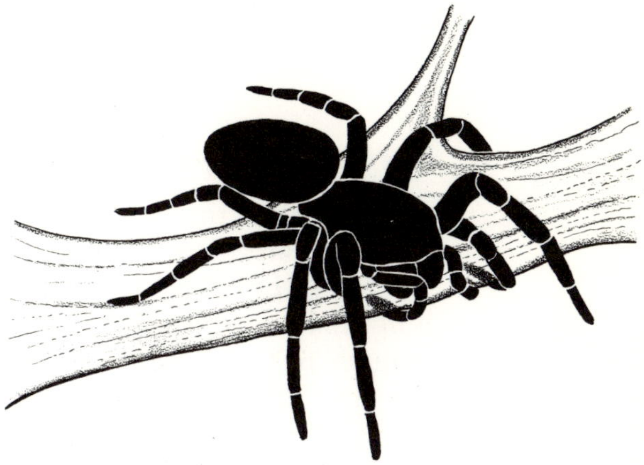

spitting spiders of the genus Scytodes shoot out a stream of sticky fluid from their fangs.

Of all the extraordinary behaviour patterns which make spiders so fascinating, those associated with mating are certainly the most bizarre. As we explained earlier, the male does not have an intromittent organ associated with the reproductive opening on the under side of the abdomen but accomplishes the act of fertilization with special structures on his palps. Before the male approaches his prospective mate he must charge these organs with sperm (Figure 13.) It is not difficult to understand why the first observers who noted this unusual method of mating concluded that there must be some internal connection in the body of the spider between the testes in the abdomen and the palp but the truth has been known for well over one hundred years. Soon after the final moult has been completed, the male spins a small web with silk from his spinnerets and many spiders then add another small sheet from the epiandrous glands above the reproductive opening. On this small sheet is deposited a small globule of semen into which is dipped the tips of the palps.

There are some variations from this general pattern, as for example, the daddy-long-legs spider *Pholcus phalangioides* which draws a single thread with the third pair of legs across the genital opening and then takes the globule up from the thread with his chelicerae from where the palps are charged. Although it would seem physically possible for some spiders to charge their palps directly from the reproductive opening, none of them do. With these preliminaries completed, the male is then ready to approach his prospective mate.

The bringing together of two carnivorous creatures can at times be hazardous and so it would be with male and female spiders if recognition devices were not used. It is always the male who seeks out the female and the urge to mate is presumed to overcome his predatory instincts for food. Nevertheless, he himself would be in constant danger if he could not in some way persuade his prospective mate that the time was opportune for matters other than eating. Male web spiders may transmit their signals in a number of ways – by plucking at the threads, by vibrating the whole body, or even by using a special stridulatory organ. The male's approach to the female is always cautious until he is near enough to touch her legs or body. The mating may then take place on the spot or the male may spin a special mating thread on to which he coaxes the female to take up the mating position characteristic of this species (Figure 14). Once the preliminary difficulties have been overcome, the female will normally remain passive as the mating proceeds and after its completion the male is able to retreat safely. Among the spiders with keener eyesight, visual recognition may play a greater part in the

Fig. 14 Characteristic stances taken during the transfer of sperm from male to female. a) *Ostearius melanopygius* Family Linyphiidae. b) *Scytodes thoracica* Family Scytodidae. Typical of all mygalomorphs and many of the six-eyed spiders. c) *Lycosa rabida* Family Lycosidae. Typical of many hunting spiders (Lycosidae, some Clubionidae, Drassidae, Agelenidae) d) *Chiracanthium* Family Clubionidae. e) *Xysticus triguttatus* Family Thomisidae.

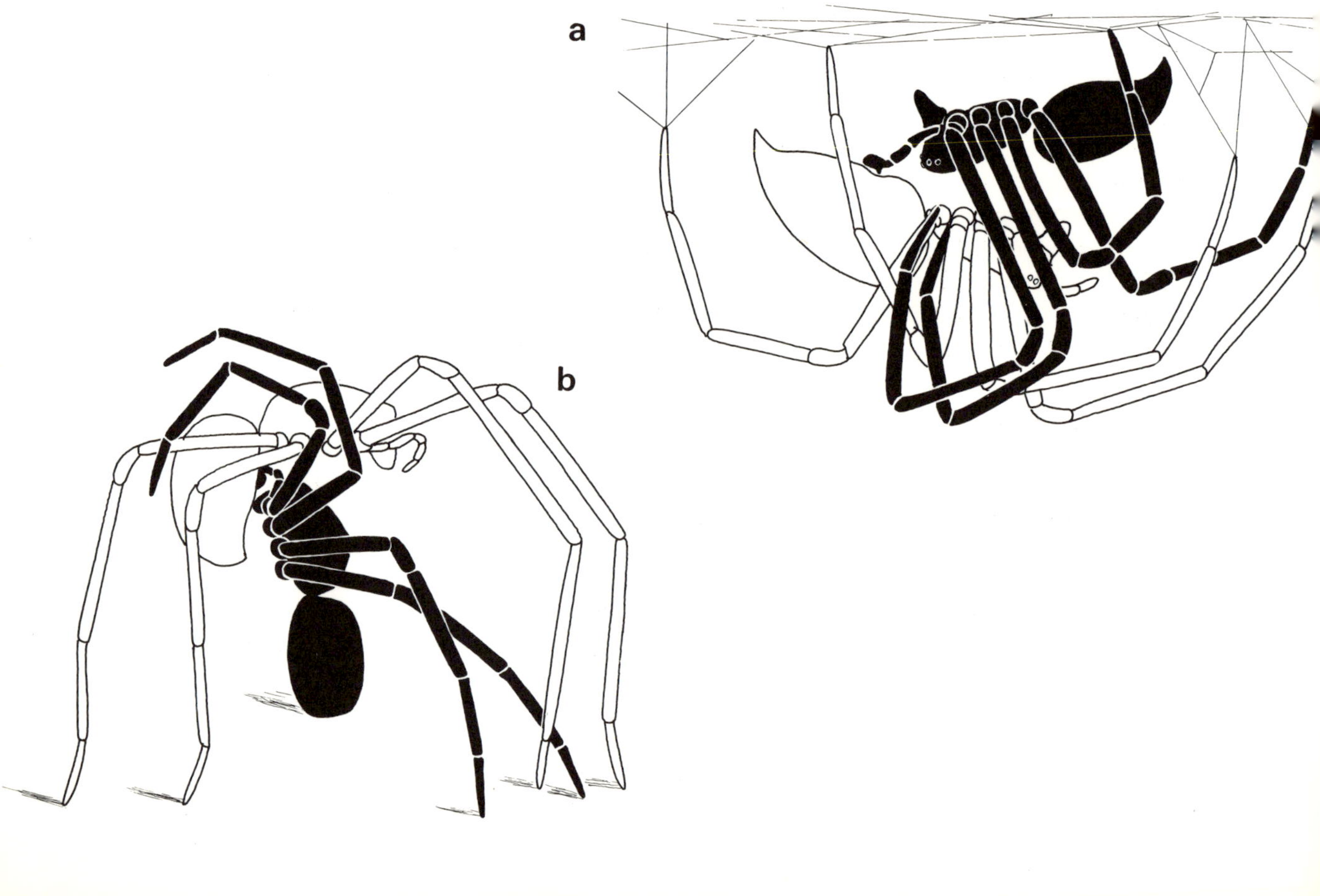

c

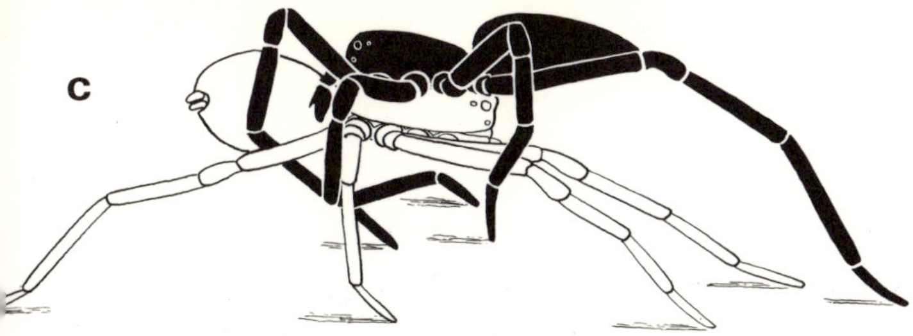

d

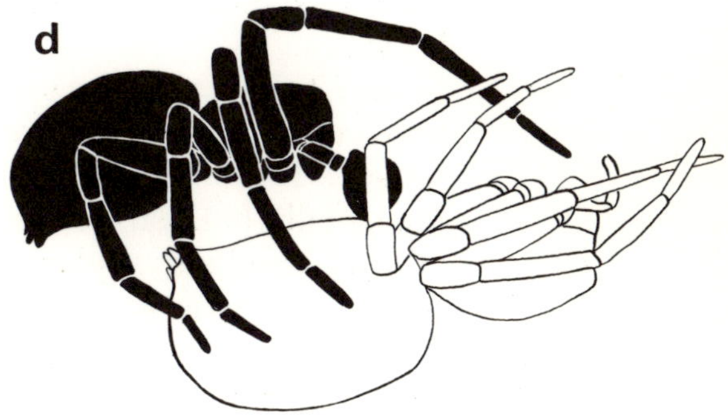

e

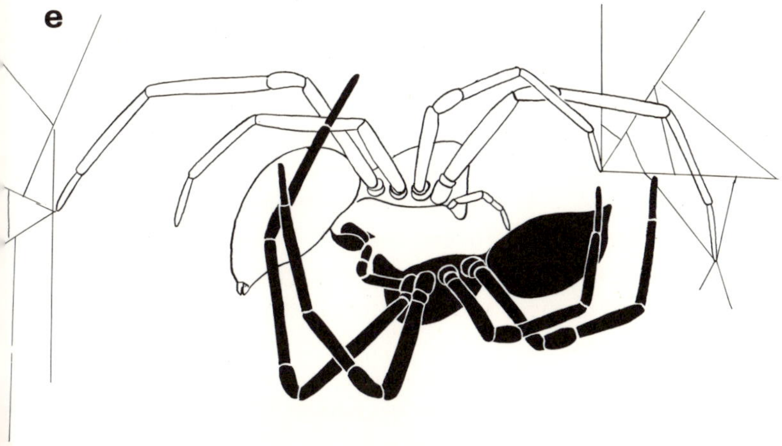

preliminary courtship and so among the hunting spiders we find a much more complex ritual. While many of the poor-sighted hunters such as the clubionids and thomisids still rely to a limited extent on vibrations, most of the recognition is achieved through touch, but as we progress through the lycosids and pisaurids to the long-sighted salticids it becomes evident that recognition becomes increasingly dependent on sight. The conclusions drawn by the pioneers in the study of this more elaborate kind of courtship were that it would enable the female to select the most vigorous and ornate performer as her mate. By selection, this would lead to a stronger stock and at the same time evolve a more complex display. These suggestions are not greatly favoured today as it has been clearly shown that the real function is recognition and possibly stimulation of the female so that her actions during mating are co-ordinated with those of the male.

Although the end result is the same in all spiders, the stance which spiders take during the actual insertion of the palpal organs varies considerably among the various groups. It is convenient to group the various stances into five main types although numerous variants are known:–

Stance 1. The two sexes are facing in opposite directions with the dorsal surface of the carapace of the male against the ventral surface of the female. The web builders such as the linyphiids which use this stance are usually both suspended upside down with the male above but in the ground dwelling spiders which assume this attitude the male stands with his body at an angle to the female.

Stance 2. In this position, the male and female are facing in opposite directions after the male has climbed over on top of the female to reach down with his palps under her abdomen. This is the attitude commonly found with the ground hunting spiders such as the wolf and jumping spiders.

Stance 3. In this position, the male is beneath the female but they are venter to venter and facing in the same direction. In the web spiders such as *Latrodectus* and *Argiope* which adopt this attitude, the spiders are suspended upside down from the web but in the crab spiders the male creeps over the body of the female and forces himself beneath her. It has been suggested by Kaston (1954) that although the end result is the same this position has been developed independently by the web and hunting spiders from stances 1 and 2.

Stance 4. The two sexes take up a position with venter to venter but facing in opposite directions. This stance has been recorded for species of the clubionid genus *Chiracanthium*.

Stance 5. This is the position taken up by the Hahniids where the male is to one side of the abdomen of the female but facing in the same direction.

Many of the terrestial spiders which use position 1 apply their palps simultaneously within the genital furrow, for these are usually spiders which have not developed an epigynum and whose palpal organ is correspondingly simple, but all the groups with an epigynum apply their palps alternately. When the palps are applied to an epigynum, it seems that the left palp is always applied to the left side opening and vice versa regardless of the position taken up during copulation. The actual copulation may be all over in a few seconds with some spiders but in others it may last for an hour or more while the embolus of the palpal organs is repeatedly inserted.

Dr W. J. Gertsch ably outlines the action of the palp in his outstanding book *American Spiders* with the following words.

The Mites – Acarina
1. Trombidiform mite from leafmould. Pohangina.

The False scorpions – Order Chelonethi
2. Te Anau, from under bark. Length 0.8 mm.

3. Female with nymphs. Port Chalmers. Length 0.5 mm.

The Harvestmen–Order Opiliones

4. *Pantopsalis* sp. Male. Lake Manapouri. Length 5 mm.

7. *Karamea lobata aurea* guarding her eggs. Oparara. Length 5 mm.

5. *Pantopsalis* sp. Female. Cascade Creek, Eglinton Valley. Length 5 mm.

8. *Soerensenella prehensor*. Female. Waipoua Forest. Note the parasitic mites. Length 4 mm.

6. *Pantopsalis* sp. Female. Waipori Gorge. Length 5 mm.

9. *Algidia chiltoni*. Male. Pohangina. Length 3 mm.

The Spiders – Order Araneae

The segmented spiders. Suborder Liphistiomorpha

10. *Heptathela kimurai*. Female, from Gifu, Japan. Showing the segmented abdomen and the large ventral spinnerets. This group of primitive spiders has not been found in New Zealand or Australia.

Trapdoor Spiders and their relatives – Suborder Mygalomorpha

11. *Porrhothele antipodiana,* Male. Family Dipluridae. Shag Valley, Otago. Length 19 mm.

12. *Porrhothele antipodiana*. Female. Family Dipluridae. Pohangina. Length 23 mm.

13. *Aparua kaituna*. Female. Family Dipluridae. Kaituna Valley. Length 7 mm.

The palpus may be scraped across the epigynum until a spur on the tibia, on the tarsus, or on the bulb itself, becomes fixed into a particular groove. Once firmly anchored in this starting point, the palpus swings to assume a position that, with the aid of ridges, grooves and other processes on the epigynum corresponding to its own outline, make it possible to guide the embolus to exactly the right point for entering the orifice. At this stage, the bulb of the palpus is still largely in its resting position, lying folded in the cup of the tarsus, and the preliminary contacts serve to hold it firmly in place. Most spiders have at the base of the bulb various thin pouches or haematodochae that swell up with the influx of blood until they attain enormous size. This distension causes the entire bulb to turn on its axis, which action forces the embolus into the appropriate opening.

Once the sperm has been deposited within the receptacles of the female, it may remain viable for a remarkably long time. Mr G. Dowling of Christchurch reports that he has kept two orbweb spiders by themselves in cages and has found that after two years of isolation they are still producing egg-sacs with fertile eggs. Unless these spiders can produce young parthenogenetically from unfertilized eggs, this must mean that the sperm has remained viable for over two years and it is now to be seen if the spiders can survive another year with similar results.

To conclude this brief survey of the life of a spider, we must consider that the main function in the life of the male has been completed with the fertilization of the female and most will shortly die, but for the female the main task is beginning because she must now fashion the egg-sac which contains the future of her species and at least one pair from the hundreds or thousands of her progeny must eventually reach maturity to ensure that the level of population is maintained.

Fig. 15 a) Mites collected from moss at Cascade Creek, Fiordland. b) Mites collected from forest leaf-litter, Masterton, Wairarapa.

CHAPTER THREE

Spider Relatives

To the uninitiated, creatures which crawl are loosely labelled as insects or bugs, unkindly and often quite falsely. For spiders and their relatives belong to the large class of animals known as the Arachnida which have four pairs of legs and a body divided into two portions, the cephalothorax and the abdomen. Insects on the other hand have bodies divided into three parts, and only three pairs of legs, and such distinctive features are easily apparent to the naked eye. A closer look at the Arachnids with a magnifying glass will also reveal that the eyes, when present, are always simple and the mouthparts, unlike many of the insects, are never complex.

While some authorities recognise as many as sixteen different groups or Orders of Arachnids at least six of these are only known as fossils, some of which date back about 200 million years to the Carboniferous Period. Of the ten remaining orders which are known to be still living on earth only three, apart from spiders, are to be found in New Zealand. These are the Order Acari or mites; the Order Pseudoscorpiones or False Scorpions and the Order Opiliones or Harvestmen. Many of the other orders, such as the true scorpions, are more at home in the Tropics; it is unlikely that they will ever be found here apart from occasional specimens which are introduced from time to time in cargo. In spite of the excitement caused when discovered they, fortunately perhaps, do not seem able to establish themselves in our climate.

All three of these arachnid groups are abundant in New Zealand and together are represented by thousands of species, the bulk of which are mites, but only the Harvestmen have been described in detail. Recently a Danish specialist in the beetle mites, Dr Marie Hammer, gathered collections of these mites during a few quick journeys throughout the country. She returned home with some three hundred different species of which only seven had been previously known to science and even this number can only be regarded as a small sample of the native fauna.

THE MITES

Mites are ubiquitous and have been found living in practically every habitat in which animals are found. While the majority are land-dwellers, some are to be found living in streams and lakes and many live in the sea. Even the frozen wastes of Antarctica have their fauna and in New Zealand mites are probably the most plentiful group of animals, both in the numbers of species and of individuals.

Understandably, the best known are those which directly affect Man; the ticks, chiggers, red spiders, cheese mites and gall mites, some of which transmit disease and others which attack crops and produce; but by far the most abundant are those free-living mites which swarm in the upper layers of the soil performing an essential role in the breakdown of organic material. A mere handful of leaf litter from the forest floor will support many hundreds of individuals and a bewildering variety of forms (Figure 15). As a group the mites are characterized by having the mouthparts, the chelicerae and pedipalps, more or less distinctly separated from the rest of the body as a false head or gnathosoma. Because the remainder of the body is fused together into one piece, they do not possess the narrow waist typical of spiders. The abdomen is smooth and not broken up into segments by regular transverse grooves, so they may be distinguished quickly in this way from the mite-like harvestmen of the suborder Cyphophthalmi. Many of the mites, however, do have a number of separate hard plates on their bodies, but this is not true segmentation.

Separation of the mites into suborders is based mainly on the number and position of the openings of the tracheal system. Six suborders are commonly accepted and most, if not all, of these groups are found in New Zealand.

As might be expected with such a large and diverse group, the life history is varied, but in general the egg hatches into a six-legged larva which is similar to the adult except that the fourth pair of legs is missing. From this larval stage, they pass through a series of eight-legged nymphal forms until they reach adulthood. While many of the nymphal forms are very similar to the adult and differ mainly in the lack of a functional reproductive system and the various associated structures, in contrast the nymphal forms of the beetle mites which dominate our soil fauna are often markedly different from the adult form.

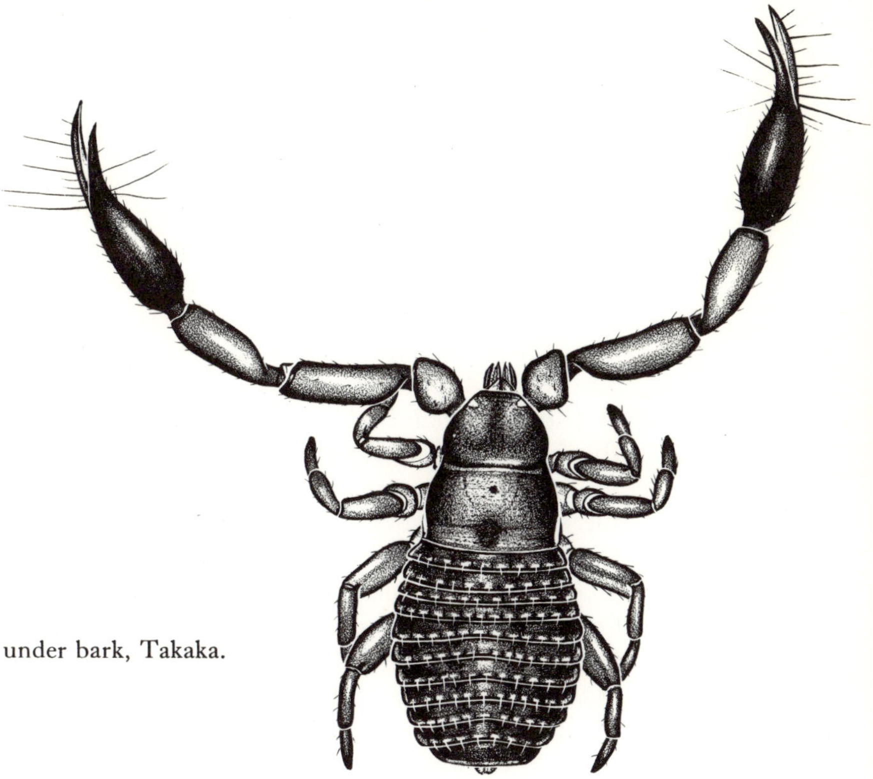

Fig. 16 False scorpion found under bark, Takaka.

THE FALSE SCORPIONS

The False scorpions, as their name suggests, seem at first glance to be baby scorpions. Over the years, many of these animals have been sent to specialists to confirm that scorpions do occur in New Zealand. Although their detailed structure differs in many ways from the scorpions, they can best be described as miniature scorpions, which, however, lack the best known portion of these animals, the stinging tail (Figure 16). They are never large and the body of our biggest species does not quite reach a quarter of an inch. Short chelicerae, with both a fixed and a movable finger forming small nippers, point out from beneath the front of the carapace. The pedipalps, however, are huge, usually longer than the body itself, and are also provided with nippers at the end. Eyes play but a small part in the life of these animals; many of them are blind, but some have one or two pairs of small simple eyes placed on the margins of the carapace.

In addition to being small, false scorpions are retiring and so, although common, they are rarely seen. Often they can be found by peeling off the loose bark from the trunks of trees and sometimes on the foliage as well, but the surest way to find them is to sift through a sieve some of the litter from the forest floor. They are by no means restricted to the forest and many species live on the seashore, some even between the tide levels, or under stones in open country or in the screes on the mountains, high above the bush-line.

Their prey is always caught alive and, while they undoubtedly eat many of the smaller forms of animal life which abound where they live, they seem to prefer those minute, wingless insects – the springtails. Hunting methods seem to be fairly standard and although, as we will mention later, they are capable of producing silk, this is not used in the capture of food.

If they are watched it will be seen that they move along slowly with the huge pedipalps probing in front; as soon as something suitable for food touches the fine hairs on the end segments it is grasped between the fingers. If, however, something too large is touched, the pedipalps are rapidly withdrawn over the body and the false scorpion scuttles backwards with the peculiar backwards movement which is so characteristic. Once an insect is grasped, it is rapidly killed with poison which oozes from one or both of the fingers and is then passed to the chelicerae. Here it is torn up; simultaneously, digestive fluid flowing from special glands in the cheliceral fingers transforms the soft parts of the prey into the liquid form it must attain before being sucked up through the mouth into the stomach.

The chelicerae do have another function quite apart from feeding; they produce silk. Only the false scorpions, spiders and mites of all the Arachnids are able to spin silk; it plays an essential part in the lives of each. In the false scorpions the silk comes from glands which open from rows of variously shaped teeth forming a peculiar comb-like structure called the serrula, attached to both fingers. Unlike the spiders, which use silk in all phases of their lives, the false scorpions use it sparingly and only for a few specific purposes. Principally it is used to fashion a retreat inside which the false scorpion may safely pass through those defenceless stages of its life – during moults, laying of eggs and hibernation.

The retreats or cocoons are usually dome shaped and, because they are usually covered with small pieces of debris, completely merge in with their surroundings. To construct the cocoon, the false scorpion uses a method which might best be compared with the traditional manner of constructing an igloo. First, a collection of fragments of debris is gathered together in the chelicerae and laid out in a ring. Then a layer of silk is spun to bind them together. As more debris and silk is added, the wall is built higher, but the width is reduced so that as the dome begins to take shape the opening gets progressively smaller until finally it is closed and the cocoon completed with the false-scorpion sealed inside. The inner surface of one of these cocoons which we examined while writing these notes was lined with a clean white sheet of silk while the outer covering consisted of small fragments of earth. This cocoon was constructed by a species which lives under loose lying stones in Central Otago; as many as twelve of these little mounds dotted the under side of every stone. Others examined on the under sides of dead leaves from the forest floor were covered with small fragments of leaf mould. But some living under stones by the seaside construct the dome of silk alone. It is usually late November when these animals are found sheltering in their cocoons while their eggs are developing.

At other times of the year, the nests protect the nymphal stages undergoing a moult or in winter would shelter hibernating adults awaiting the coming of warmer spring weather.

Perhaps before we describe how the female guards the eggs during their development, we should outline some of the bizarre events which took place earlier during the courting period and the method by which she was fertilized. Only fragmentary information is available on the behaviour of our native species, but what little we have seen shows that the extraordinarily detailed observation made on these animals in other countries, and particularly the studies of Dr Max Vachon of the National Museum of Natural History of Paris apply in greater part to our own fauna. Courtship, as with the spiders, is initiated by the male who, as he slowly approaches the female, vibrates his body and moves his pedipalps backwards, forwards, and upwards in a series of regular postures. At the same time some species extrude from the under side of the abdomen a pair of long horns which extend like the fingers of a glove from beneath the front of the body. During all this activity on the part of the male, the female, if she is receptive, remains passive. When the male is near, he lowers his body to the ground and as it is raised again a slender thread of viscous material is drawn out from the reproductive opening beneath his abdomen. This thread shortly hardens into an erect rod firmly fixed to the ground. On the tip of this structure which is called the spermatophore or sperm carrier, he places the drop of seminal fluid required to fertilize the eggs, but then it is necessary to ensure that this minute ball of sperm is safely deposited within the egg duct of the female. To do this a number of methods are used, each of which is characteristic of a particular species. In some species the male, after moving back from the spermatophore, will vibrate his pedipalps and at this signal the female will slowly advance until she is straddling the rod. The male now advances and grasping the base of her pedipalps in the fingers of his, shakes her, and in doing so the spermatophore is inserted and the drop dislodged. Other species use a more direct method where the male moves forward and by grasping the pedipalps of the female draws her back over the spermatophore; while in some species, the pedipalps are held both during courtship and the actual impregnation.

As soon as her eggs begin to mature the female constructs the dome retreat in which she will remain until her young have hatched and developed sufficiently to fend for themselves. Before the eggs are actually laid, however, further preparations must be made. From special glands in the egg duct a fluid is excreted which bulges out from the opening and hardens to form a small sac. This is the incubation chamber and it is inside this structure that the eggs are laid, each being fertilized as it passes down the egg duct. The eggs hatch shortly into a larval stage which shows no resemblance to the adult animal as it is little more than a small white bag with only one pair of appendages – the pedipalps, and a short, muscular beak. These helpless larvae represent a much earlier stage of development than most other arachnids when they hatch from the egg and as they are without a store of food to see them through to the next moult they are directly dependent on the mother for nourishment. This she provides in a most unusual way. Her ovaries break down and the resulting yolk floods down to fill the incubation chamber where it is sucked up through the muscular beak of the larva. It is as if this

Fig. 17 False scorpion mother with distended larvae, in retreat. Note how the abdomen has flattened and is pushed high above the carapace.

method is too slow because a few days later a sudden muscular contraction of the mother's abdomen will force a vast amount of food into the bodies of the larvae so that within a few seconds they will have swollen to as much as three times their previous size. The incubation chamber also expands with the result that the mother's abdomen which is now flattened after the transfer of the food, is pushed high up into the air (Figure 17). Over the next two weeks this enormous meal will be absorbed and the typical form of the false scorpion slowly develops. When it then moults into the first nymphal stage, it is able to burst first from the incubation chamber and then from the silken retreat, to make its own way in life.

THE HARVESTMEN

Of all the arachnids, the harvestmen epitomize the unusual and grotesque, but they are in fact the most harmless of creatures and one of the few which have no use for poison. Curiously enough, although native harvestmen are numerous and widely spread, the first type most people encounter in this country is usually the common brown European harvestmen *Phalangium opilio* (Figure 18). This is the harvestman seen running about in gardens and open country throughout the summer months and is the only harvestman which is found from one end of the country to the other. Because this species is also numerous in the fields in England during the harvest period, it has gained for the whole group the common name of harvestmen. To see the compact oval body racing through the air suspended from eight tremendously long and extraordinarily slender legs is an astounding spectacle in itself, but this species is one of the most conservative of a group which specializes in the development of seemingly useless ornamentation and modifications.

As well as the usually accepted name of harvestmen, the Opiliones are sometimes called daddy-long-legs – a name more commonly and better used for the Crane-flies. Some of the characteristics which separate the harvestmen from the spiders are negative ones, such as the absence of silk-producing organs and poison glands, but the most obvious, and recognisable feature is the compact body which is not divided into two parts as in the spiders. Some of the harvestmen are mite-like, but these species always have grooves across the abdomen separating the segments whereas in the mites the abdomen is smooth. Some are blind, but the majority have two simple eyes placed on each side of a mound rising like a look-out from the middle of the carapace. The chelicerae are always

Fig. 18 The garden harvestman *Phalangium opilio*. Introduced many years ago from Europe, it is now found throughout New Zealand.

chelate with the nippers at the end of the second segment and the pedipalps are leglike although in the Laniatores they are large and armed with spines.

The term scavenger seems most appropriate to describe their feeding habits, as most will eat any dead invertebrate which they come across. Many of them also prey upon small insects, the long-legged harvestmen grasping them with their chelicerae and the short-legged harvestmen with their formidable spined palps. These insects are located by the ends of the long second pair of legs which are always probing in front and to the side as the animal moves along. In captivity they have thrived on an extraordinary range of diet, from pieces of meat and dead insects to bread, fat and fruit juice. In the absence of poison glands, it might be thought that these animals are easy prey to their more formidably armed neighbours, but after a brief encounter these hopeful predators quickly back away, repelled by a squirt of acrid fluid from the small openings on each side of the carapace which are the vents for stink glands. This fluid alone seems to ensure a safe passage, as even the most pugnacious of the spiders avoid them. The harvestmen are mainly nocturnal creatures, which hide in the dark during daylight; because they easily dry out and die they are found in places where the humidity is high. For this reason they are mainly forest dwellers and to find them in the daytime you would look among the leaf litter or under logs and stones on the forest floor. A few do live outside the forest among tussock or under loose stones on the mountainsides and riverbeds while some live only in the permanent darkness of deep caves.

Unlike many arachnids the harvestmen are most undemonstrative before mating and no true pre-nuptial courting behaviour has been recorded. The mating of our long-legged harvestmen has never been observed, but that of the common *Phalangium opilio* consists of little more than the male taking up a position in front of the female when fertilization is achieved by the extension of a long intromittent organ from a pouch within the abdomen of the male. With the short-legged harvestmen much the same thing happens but as the intromittent organ of the male is relatively shorter, the male takes hold of the palp of the female with his own palp and then raises both of their bodies so that the openings on the under side of the abdomen are closer before the actual mating can be accomplished.

The mating behaviour of the mite-like harvestmen of the suborder Cyphophthalmi is completely unknown, but from anatomical studies we know that this must be quite different; in fact it is doubtful whether these animals should be included with the harvestmen at all. The internal structures show that the male produces a spermatophore so that the actual mating will be indirect as in the false scorpions. The details of what really does happen should be most interesting. It is to be hoped that these facts will be elucidated by some student in this country where almost as many species have been recorded as in the rest of the world put together. The eggs are laid through a long ovipositor, usually in small groups in some moist dark place such as amongst the leaf-litter or moss or below stones and fallen logs. The young which hatch from the eggs three weeks later are similar to the adults, but colourless, and are able to fend for themselves within a few days.

The harvestmen are separated into three suborders which are easily recognised.

1.) Cyphophthalmi – the mite-like harvestmen.
2.) Laniatores – the short-legged harvestmen.
3.) Palpatores – the long-legged harvestmen.

The Cyphophthalmi, in contrast to their name, are small animals and the body of the largest species in New Zealand does not exceed five millimetres in length (Figure 19). These drab, dark-brown harvestmen live a hidden life among the leaf-litter and moss on the forest floor where it is presumed they feed on small insects such as springtails. All the New Zealand species are blind and the mounds on each side of the carapace which might be mistaken for eyes actually house the stink glands. The sexes are alike except for two striking modifications in the male where the end segments of the abdomen are depressed in the centre so that each side shows up as a hump and the tarsus of the fourth leg is provided with a stout spur from which empty a number of special glands clustered like a bunch of grapes inside the tarsus itself. The purpose of these structures, which are different in each species, is not known but they probably play some part in the mating behaviour. The two genera found in New Zealand are easily separated in the males because the fourth tarsus is a single segment in *Rakaia* while it is divided into two in *Neopurcellia*.

The Laniatores are the best known group in New Zealand and about one hundred and sixty species are recorded. These are the harvestmen whose profusion of seemingly useless spines, knobs and pustules could raise some doubts as to the advantages of evolution within this group. Their common collective name is 'short-legged' harvestmen although it must be admitted that this name can be rather puzzling because none of them have particularly short legs. In fact some of them are quite long-legged. To be precise, they should perhaps be called the 'shorter-legged' harvestmen to separate them from their extraordinarily lanky relatives – the Palpatores.

Characteristically the Laniatores have extremely stout pedipalps armed with spines for use as grasping arms. A hard hinged plate which closes the opening to the internal pouch, just behind the legs, holds the reproductive organs concealed within the abdomen on the under surface. By contrast, the Palpatores usually have smooth and slender pedipalps in addition to long and slender legs while the reproductive opening is covered with a soft flap.

When the New Zealand Laniatores were first examined in detail a number of years ago, it was found that in many species two quite different forms of male were associated with one kind of female. One of these males would be very much like the female, but the other, as if flaunting his manliness, had much longer and stouter pedipalps and longer chelicerae. The reason for these two different males existing is still obscure, particularly as experiments show that as far as the female is concerned she is not particularly impressed by this show of strength and will mate with either form which happens along at the right time.

equivalent genus in the North Island *Soerensenella*, has only a small erect spine on the eyemound (Figure 22). The behaviour of all of the species of these two genera is similar. As the eggs begin to mature within the ovaries, the female harvestman ceases to feed and settles down in the darkness beneath a log or stone. It is at this time that the male seeks her out and mating takes place. Before long, she deposits a small patch of from twenty to thirty eggs on the surface beneath her and remains crouched over them like a broody hen. In normal circumstances these eggs would hatch out in three weeks and her vigil would be over, but in a few days time she lays a few more eggs and at regular intervals continues to lay even more. The eggs are fertilized only as they pass down the eggduct to be laid and by the time the first batch is hatching there are a number of other batches at all stages of development. She does not feed while she is guarding the eggs and so it might be thought that she carries on till she drops dead of starvation, but in fact the emergence of the first batch of young signals the end of egglaying. Thus her vigil is for six weeks or just twice as long as it would have been if she had laid all her eggs at once.

Before completing this brief survey of the short-legged harvestmen mention should be made of the pygmies of the group which again have a name in direct contrast to their size. They belong to the family Synthetonychidae and are all placed in the one genus *Synthetonychia* (Figure 23). The body of the largest of these animals is less than two millimetres and they are found only in the damp moss of New Zealand forests. Apart from their small size, these harvestmen can be recognised by their compact body, the strong smooth pedipalps and the absence of

Fig. 22 *Soerensenella rotara* Family Triaenonychidae. During the summer months these harvestmen guard a batch of eggs laid beneath logs in the forest.

Fig. 23 *Synthetonychia* sp. Family Synthetonychidae. These minute harvestmen, found only in New Zealand, live amongst moss and leaf debris in the forest.

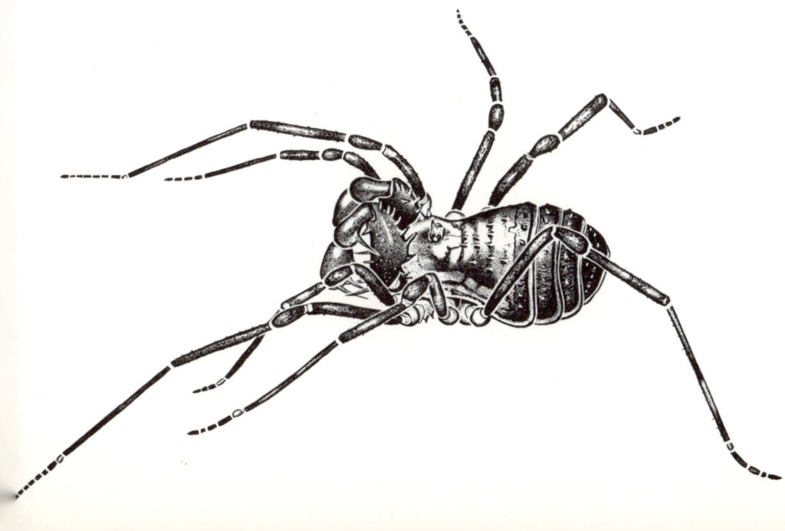

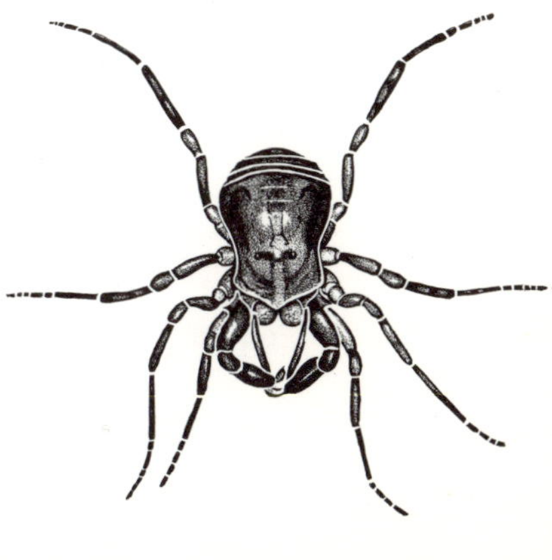

an eyemound, with the result that the eyes are perched flat on the middle of the carapace. About the only way of finding these animals is by sifting through moss and as no one has yet studied the live animals, nothing is known about them other than that they are unique to New Zealand.

Quite the most comical of all the arachnids are the Palpatores or long-legged harvestmen. It just doesn't seem possible that the long threadlike legs only a millimetre thick and perhaps twenty times the body length could support the animal to which they are attached yet in fact they not only manage, but manage extremely well. In spite of their ungainly appearance they are remarkably agile and can put on a surprising burst of speed when disturbed; what is more, they are still quite mobile when only three or four legs are left. Many of these harvestmen may be seen with missing legs which are always broken off at the end of the first segment near the body. At this point the joint is especially weak so that the slightest pull on the leg will cause it to break; thus the leg remains while the harvestmen escapes. With this useful built-in escape mechanism it is surprising to find that the harvestmen are one of the few groups of arachnids which are not able to regenerate a limb that has been lost.

In most of the long-legged harvestmen from other countries, there is little difference between the sexes, but in many of the New Zealand species the situation is different. The males and females of species of the common genera *Megalopsalis* and *Pantopsalis* are so dissimilar that for many years they were thought to be distinct species (Figure 24). In these genera the male is usually black sometimes relieved by red or yellow markings and the carapace is covered with minute spines. By far the most striking characteristic, however, is the huge and ungainly chelicerae possessing only two segments, but towering high above the body like a huge crane. In contrast, the female is a most attractive creature with a smooth compact body, beautifully coloured with bands and patches of yellow-brown, green or red while the chelicerae are so small they hardly show above the carapace. Of the hundreds of species which live in this country, only a dozen or so are described and little is known of their habits, but it is thought that the chelicerae of the male perform some role in mating behaviour.

Zeopsopilio novaezealandiae is a minute member of the Palpatores found living in leaf-litter which makes up for its small size by its grotesque form. The eyemound is huge and the eyes correspondingly large. The pedipalps are intermediate between those of the long and short-legged harvestmen, being armed with spines and used to capture their minute

14. Trapdoor nests of young *Cantuaria vellosa*. Family Ctenizidae. There are five nests present.

15. Open-burrow Trapdoor Spider, *Cantuaria borealis*. Female. Family Ctenizidae. Kaituna Valley. Ground-dwelling mygalomorph spiders are often attacked by parasitic mites which cling to the carapace where the legs cannot reach. Length 11 mm.

16. Lidded Trapdoor Spider, *Cantuaria assimilis*. Female. Family Ctenizidae. Enfield. Length 30 mm.

17. *Cantuaria assimilis*. Male. Family Ctenizidae. Enfield. Length 13 mm.

18. Tree Trapdoor Spider, *Migas goyeni*. Female. Family Migidae. Kaituna Valley. Length 12 mm.

19. Nest of *Migas goyeni* on a tree fern trunk.

The Four-lunged true spiders Hypochilomorpha

20. *Gradungula sorenseni*. Female. Family Gradungulidae. Knights Point, near Haast. Length 7 mm.

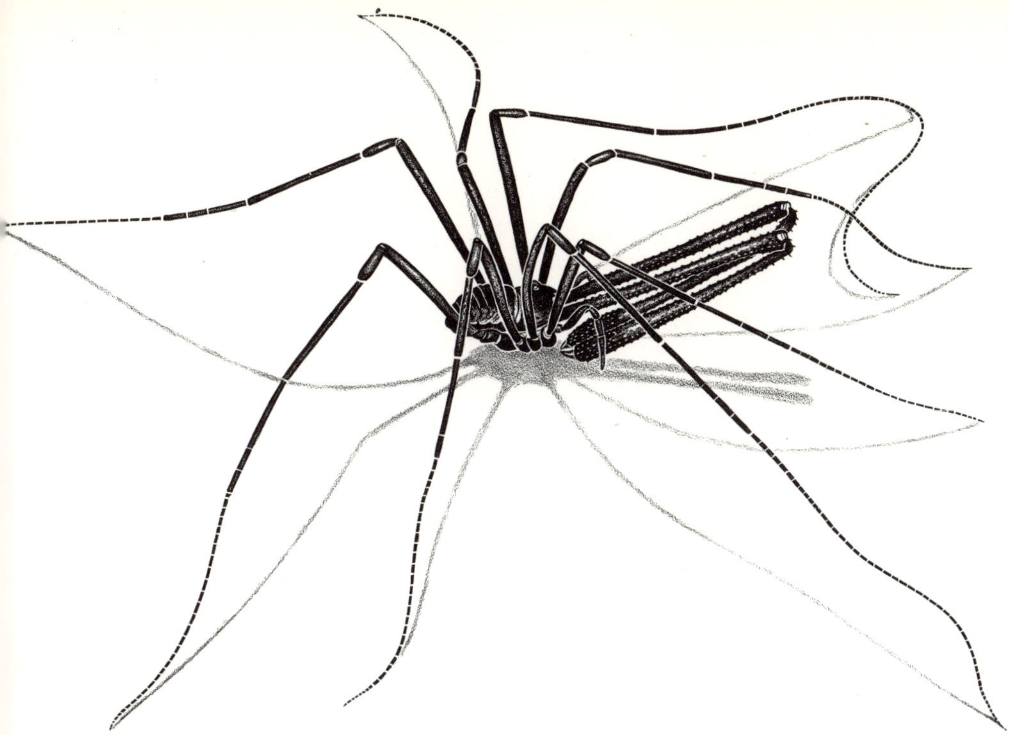

Fig. 24 *Pantopsalis* sp. Family Phalangiidae. These familiar long legged harvestmen live mainly in the forest but are also found in open country including the sub-alpine region. The male (a) and female (b) are so different that for many years it was thought that they belonged to different genera.

prey. Today these harvestmen are extremely rare and have been found only in New Zealand, Australia, South Africa, the southern regions of South America and a few places in the United States. Incredibly, they have also been found beautifully preserved as fossils in amber from the Baltic of Europe where they were trapped in the gum oozing from trees growing some 20 million years ago. It seems that at some time in the distant past these little animals enjoyed a much wider distribution than they do today.

PART II *The Groupings of Spiders*

It is not possible at present to discuss the New Zealand spiders on the basis of a selection of those species which will be more commonly found because the majority of them, including many you will find in your backyard, have not yet undergone the formal phase of description in a scientific journal which is the necessary preliminary to the attachment of a proper and permanent name. Only some five hundred species of the probable total of 2,500 which live in this country have as yet been described. As these circumstances apply equally well to the grouping of species into genera, the aim in the following chapters is to facilitate recognition of the different kinds of spiders, first at family level and then into genera and species where these are known.

It might be thought that after some two hundred years of serious study at least the broader grouping of spiders into families would be settled but unfortunately this is not so. Some authorities consider that familial grouping should be established mainly on the basis of convenience, so that those animals which look alike are placed together regardless of the fact that they may have evolved from quite different ancestors and become secondarily similar. Doubtless this method has many advantages if the sole purpose of a classification is to place a specimen in a convenient pigeon-hole, but it is generally accepted nowadays that the classification of an animal group should also throw some light on the true relationship of the living forms and if possible the direction which evolution has taken during the development of the group through the geological past. This needs to be said because the classification of the spiders is today in a state of flux as the change is under way from a system based mainly on superficial similarities to one based on what is judged to be true ancestry. In the survey of the New Zealand representatives which follows we have been conservative, or as it might be said old-fashioned, and have used in general the groupings and names which are found in most of the standard works in preference to those which would need to be searched for in specialized literature. Some changes will, however, be made shortly in the taxonomic revision of the New Zealand spiders, which is at present being published, so mention of possible changes has been made in the text to provide a lead for those who may wish to study these fascinating animals seriously.

CHAPTER FOUR

The Mygalomorph Spiders

The Mygalomorphs or trapdoor spiders and their kin, are easily recognised by their large size and solid build, the up-and-down movement of the fangs when they capture prey and also by the presence of four book lungs on the under side of the abdomen (Figure 5a). The unique articulation of the fangs can be seen easily enough if the spider is turned over, as they lie more or less parallel to each other when at rest. If the spider is alive and can be made to strike at a stick brought near, it will be seen that these spiders must first raise the head by rearing up to provide clearance for the downward strike of the fangs (Figure 25). This action is in strong contrast to the behaviour of other spiders with transverse fangs which are able to bite without going through this procedure. The four book lungs usually show up clearly on the under surface of the abdomen as pale patches, one pair near the waist and the other pair more or less in the middle.

Three families of these spiders are found in New Zealand. The Ctenizidae are the ground-dwelling trapdoor spiders, the Migidae are the tree trapdoor spiders and the Dipluridae are the tunnelweb spiders. Although the habits of individual species of each of these three families fall nicely within these three categories of behaviour and the places where they live, there are exceptions. Some of the Migids construct their trapdoor tunnels in the ground and then these look like the home of the young of true trapdoor spiders. Not all of the true trapdoor spiders close the openings of their burrow with a lid and, to add further confusion, some of the Diplurids have developed the habit of constructing burrows which in appearance are identical with those of these exceptional trapdoor spiders. Fortunately, however, all the spiders themselves are very distinctive and it is easy to place them in the correct family.

The Diplurids are the largest of our mygalomorph spiders and, at least by weight, are the largest of all our spiders. The body length of some

Fig. 25 A male of *Porrhothele antipodiana* takes up an aggressive stance.

Fig. 26 Female *Porrhothele antipodiana* Family Dipluridae. The long superior pair of spinnerets projecting out from beneath the abdomen provide the silk used to construct the tunnel retreat.

may exceed an inch. Some have the unfortunate habit of wandering into the house at night where they become trapped in the bath or sink, causing consternation in the household in the morning. These are invariably males which have been wandering about in search of a mate. Three genera are found in New Zealand and all of them may be recognized by the extremely long posterior pair of spinnerets which project back beyond the abdomen like two small, flexible fingers. In spite of their size and the fact that they are related to the notorious tunnelweb spider of Sydney, very few people get bitten and even those few victims have suffered little more ill effect than a slight inflammation at the site of the bite.

The best known Diplurids are those of the genus *Porrhothele* – large spiders with a bright yellowish or orange carapace and a bluish-black abdomen. They construct a bulky silken tunnel or tube beneath a loose fitting stone or log, within a hollow tree, or in the crevices of a broken rock face. The commonest species is *Porrhothele antipodiana* (Figures 25, 26) whose tubes may be ten inches long and an inch in diameter, opening out on to a sheet of silk which probably acts as a signalling device to warn the spider when its next meal is approaching. Once an insect is captured, it is quickly dragged back into the tunnel to be eaten. The remains of these meals can be found, neatly sealed off with silk at the end of the burrow and so it is not difficult to discover the spider's favourite diet. Although it seems that these spiders will attack and eat most arthropods, beetles always form the bulk of their prey.

Despite the fact that these spiders are widespread, being found all over the North Island, and the South Island, north of Balclutha on the east coast and Hokotika on the west coast, no-one seems to have observed their full mating behaviour. While we were not successful in seeing the final stages, we were able to watch a male as he approached a female. As he touched the silken platform he raised his front pair of legs so that the peculiarly excavated portion of the metatarsus and the swollen and spiny section of the tibia were directed to the female and at the same time he lifted up the front portion of his cephalothorax to expose his fangs and palps. With the front legs vibrating, he moved slowly forward until he reached her but unfortunately the necessary responses from the female did not eventuate and so they remained head to head for some twenty minutes and nothing else took place. It is to be presumed that the male would grasp the female in such a way so that her fangs are held between the stout tibial bulge and the excavated metatarsus while he inserted the tips of the simple bulb-like palpal organs to complete the fertilization. To do this successfully, it is probable that she would need to raise the front portion of her body so that he could reach her fangs with the modifications of his front legs and then hold her up while he reached forward with his palps to her reproductive opening on the underside of her abdomen. This head to head mating procedure is common to most of the mygalomorphs.

In early or middle summer the female constructs a single bulky egg-sac containing from two to three hundred loosely wrapped eggs which she attaches to the wall at the end of the tube. The spiderlings hatch in thirty days but remain with the mother in her tube for some time before leaving to spin their own similar, but miniature, homes nearby. Little is known of the lifespan of these spiders but it is probable that two or three years pass before maturity is reached. If the spider is a male, then only a few months of life remain but for the female maturity does not mean that her life is drawing to a close because she can continue to moult and grow. Some females have been kept in captivity for four years after they have reared their first brood and so at least six years can be suggested as their possible life span.

Fig. 27 *Hexathele hochstetteri* lives in a loose silken tunnel beneath stones and logs or in crevices.

Fig. 28 *Hexathele maitaia* within her retreat.

Many of the species of *Hexathele* look superficially like those of *Porrhothele* and the only way to be sure of identifying them is to count the number of spinnerets. *Porrhothele* has only four, but *Hexathele*, as its name suggests, possesses an additional pair to bring the total to six. Although six spinnerets is the normal number for the true spiders, very few mygalomorphs have more than four and *Hexathele* is the only genus found in New Zealand with six. The commonest species of *Hexathele* in the North Island are large, with a yellowish-brown carapace and a prominent chevron pattern on the abdomen. These spiders belong to *Hexathele hochstetteri* (Figure 27) or related species and all of them construct a loose silken tube under stones or logs in much the same way as *Porrhothele*. A number of dark, sometimes black, species is also found in the North Island; some, such as *H. maitaia* live in typical surface tubes under logs (Figure 28), but others live in burrows in the ground. Typical of these subterranean species is *H. puka* from the Wairarapa, which excavates a deep burrow, lightly lined with silk from which a few radiating threads probably act as warning triplines. The South Island species of this genus range from some which still show

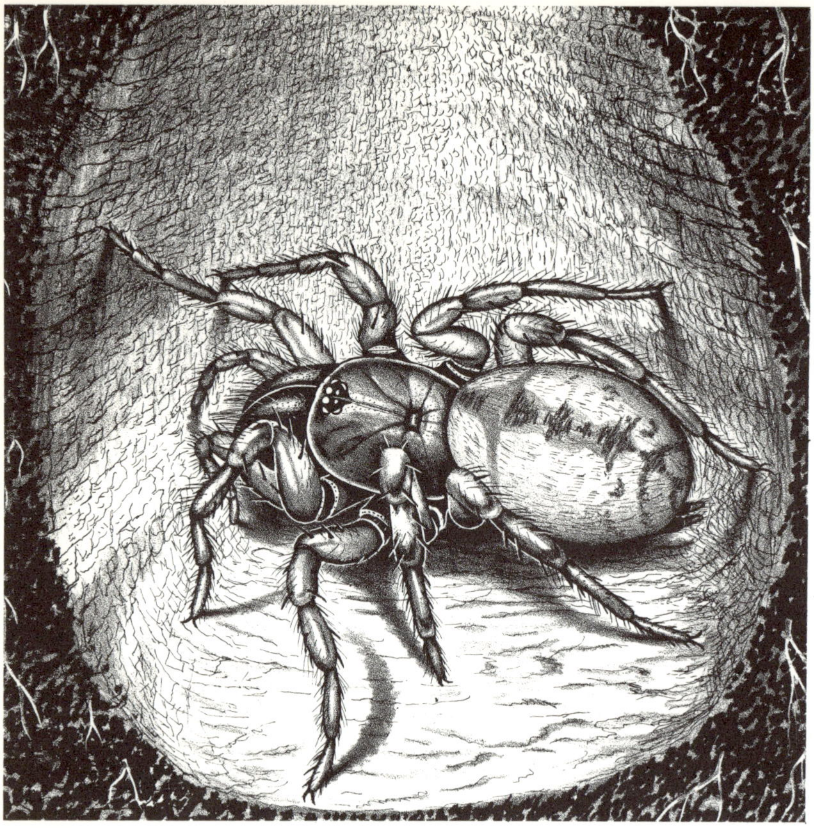

Fig. 29 *Hexathele petriei* from Central Otago rests in her tunnel in the ground.

traces of the chevron pattern on the abdomen to others where the chevrons have joined until they can barely be seen, and again some live in silken tubes while others live in burrows. Two strikingly different yellowish brown species are found in the South Island; in these spiders, the chevrons have been reduced to faint remnants on the hind portion of the abdomen. One of these species, *H. petriei* (Figure 29), from inland Otago, is found in the dry open country typical of this area, where it lives in tunnels in the ground (Figure 30). P. Goyen, one of the early school inspectors in Otago, who first described this species and gave an account of its habits in 1886, suggested that the spiders left their burrows to forage for prey, but this is doubtful because pitfall traps set near a concentration of burrows trapped only mature males, suggesting that like most mygalomorphs these spiders capture only those insects which come within reach. Surprisingly the closely related *H. rupicola* which lives in the riverbeds of Canterbury shows no tendency to burrow. but constructs a typical silken tube amongst the stones which accumulate near the banks.

Many of the spiders placed in *Porrhothele* and *Hexathele* have been known for a long time. In fact *Porrhothele antipodiana* has the pride of

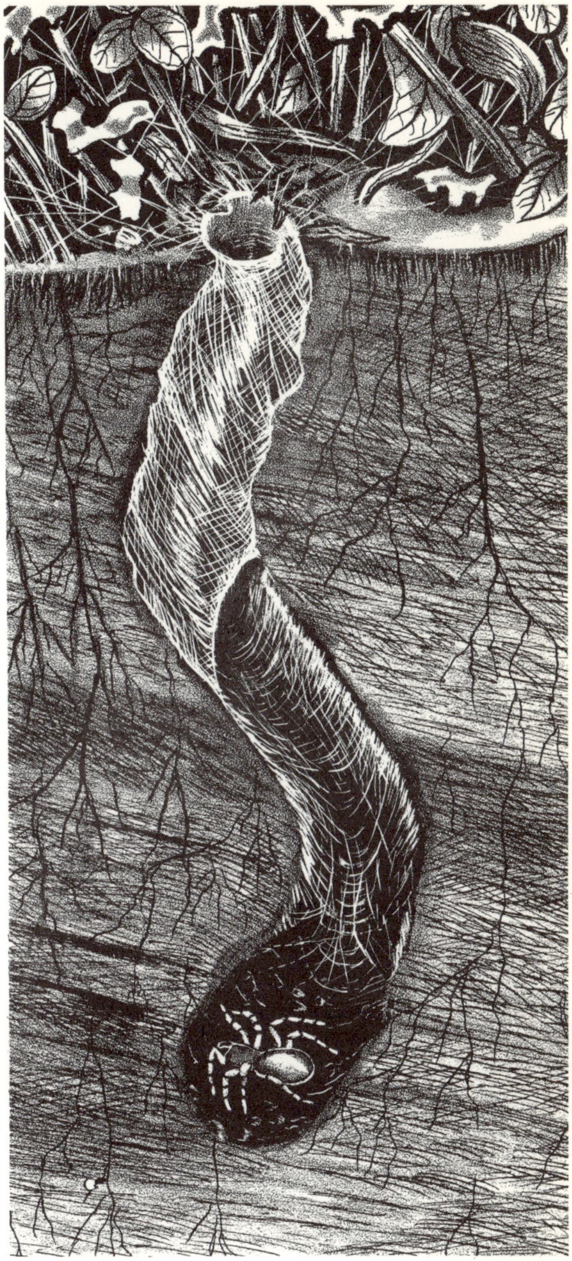

Fig. 30 *Hexathele petriei* a) The open entrance to the tunnel. b) Section of the tunnel.

place among our spiders as being one of the first ever to be collected in New Zealand. It was first described in 1849 from specimens which had been collected in 1827 by the French naturalists Quoy and Gaimard during the voyage of the French corvette *Astrolabe*. *Hexathele* was established in 1871 by the authority on the mygalomorph spiders at that time, the Australian Anton Ausserer. He named the species he examined, *Hexathele hochstetteri* after his fellow countryman Ferdinand von Hochstetter who carried out much of the early geological exploration of New Zealand.

While a number of other diplurids had been discovered over the intervening years they all belonged to either *Porrhothele* or *Hexathele* and it was thought that no radically different form of these large spiders could have been overlooked in such a small and comparatively well populated country as New Zealand. It was therefore a great surprise when in 1944 Miss Valerie Todd, who was studying this group of spiders for her Master of Science Thesis, found an entirely different species living in open-mouthed burrows in the forest floor near Wanganui. This species she placed in a new genus called *Aparua* from the Maori word meaning a hole (Figure 31). At first glance these spiders seem to be true trapdoor spiders. The abdomen is mottled brown and they are squat in appearance, but like all diplurids the posterior pair of spinnerets is long. Once the presence of these spiders became known they were found to be quite common. We know that different species may be found from the Three Kings Islands in the far north to Stewart Island in the south. The egg-sac is found attached to the wall at the bottom of the burrow with the female, during the spring months. When the spiderlings emerge they do not seem anxious to leave and continue to share the nest with the mother for some time. This reluctance to leave may even extend over a full year because sometimes tunnels are found still occupied by half grown spiders of the previous year's brood while the mother spider is waiting for the next year's eggs to hatch.

Fig. 31 *Aparua kaituna* Family Dipluridae. Most of the species of this genus are found in forest. This species from Banks Peninsula lives in a deep open burrow in the forest floor.

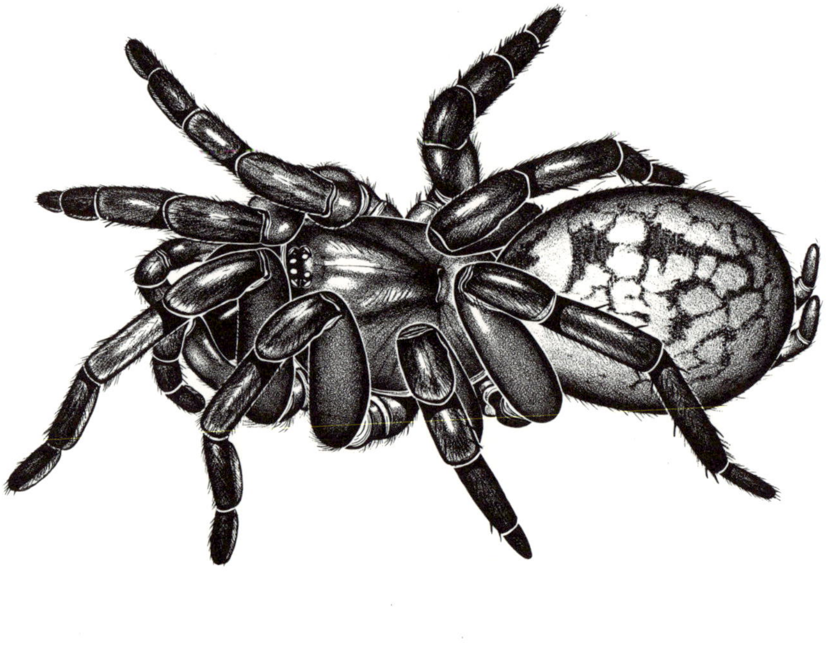

Fig. 32 The open burrow of *Cantuaria huttoni* is similar to that of some of the species of *Aparua* and *Hexathele*.

The true trapdoor spiders of the family Ctenizidae are abundantly represented in New Zealand, particularly in the South Island. At present the forty species described have all been placed in the New Zealand genus *Cantuaria*. The majority of these spiders close their burrows with a thin lid which rests on the top of the hole although a number of forest species leave the burrow open (Figure 32). This type of lid is called a 'wafer lid' to distinguish it from the thick 'plug lid' which some overseas representatives of this family construct. All the spiders of this family are characterized by the group of strong spines on the front face of the chelicerae called the rastellum. The four spinnerets are small and cannot normally be seen from above.

During the excavation of the burrow, the palps, the first pair of legs and the chelicerae are used to scrape the earth away. The grains of dirt are held between the fang and the base of the chelicera, brought to the surface and dropped a short distance away from the opening. It is in fact these small mounds of discarded earth appearing at intervals as the spider enlarges its burrow which provide the clue to their presence. Nevertheless even when it is known that the opening is in the immediate vicinity it is still most difficult to see. As the burrow is excavated, the walls are firmed by pressure from the fangs but the internal silken lining is not spun until the tunnel is complete. When this stage is

Fig. 33 The trapdoor tunnel of *Cantuaria toddae* from Central Otago has a wafer type lid.

reached, the burrow lining covers the opening, enclosing it completely. Subsequently two-thirds of the margin is bitten through, leaving the other third as a hinge. With this type of hinge the door will not flip back and will shut as soon as it is released. This would be a disadvantage to the spider if it were locked out of its home but providentially these spiders rarely venture out. They lurk at the top of the burrow just below the lid with the door slightly open until some insect ventures near, perhaps even actually touching the lid. Almost casually the spider leans out to capture it and drag it back into the burrow. Immature males behave in the same way but, like most spiders at maturity, they abandon their homes, and set out to find a mate, and at least some of them do not eat again. It is rather surprising to find that the males mature during the winter months, for it is from April to June or July when they are found wandering about in the open looking for the burrows of their prospective mates. The full mating behaviour has not been described for any one species but from random observations it appears that their behaviour is fairly uniform. The male signals to the female by tapping the lid of her burrow with his front pair of legs which are held straight out in front of him, and at intervals vibrate rapidly. It is not clear whether a responsive female then opens the lid to let him in or whether in the absence of any adverse reaction he opens it himself but the actual mating is always accomplished within the tunnel.

Fig. 34 *Cantuaria huttoni* female. Family Ctenizidae is typical of the forest dwelling trapdoor spiders.

Fig. 35 *Cantuaria napua* from near Palmerston constructs its lidded burrow in open grassland.

The identification of trapdoor spiders to species is a difficult task but there is one group which is easily recognized. This is the forest species, *Cantuaria huttoni* (Figure 34) and its relatives, which live in open mouthed burrows on the forest floor. Most of these spiders have a prominent chevron pattern on the abdomen. Once the eye has become accustomed to seeing the half-inch wide silk-lined openings on the forest floor, they can be found without difficulty but it is not always a simple matter to secure the spider. At the slightest disturbance, it retreats to the bottom of the tube which wends down some eight or nine inches between stones and roots. The silk lining becomes thinner as the bottom of the tube is reached and so before digging it up it is wise to push down gently a long straw to act as a marker as the tunnel is carefully excavated. When placed in a container of earth the spider will immediately begin to burrow and the methods of excavation can easily be seen.

The typical trapdoor spiders are found in the open country where they often occur in vast numbers but they are so retiring that the local residents are quite unaware of their presence. Perhaps the greatest concentration is in the Palmerston-Otago district of the South Island where a score or more species may be found, each restricted to a small area although two or three species may live together, even in adjacent burrows. Miss Lyndsay Wilkinson who is studying the common species *Cantuaria napua* near Palmerston has found that an acre of undisturbed land may be the home for many thousands of these spiders, (Figure 35).

The name 'tree trapdoor spiders' is usually used for the spiders of the Family Migidae although by no means all of them live only on the

Fig. 36 The males of *Cantuaria* have processes on the tibiae of the front pairs of legs used to hold the female during mating.

trunks of trees. They are generally more squat and smaller than the true trapdoor spiders, being mostly dark brown or blackish in colour. When constructed on tree trunks the nests are very hard to see and we find that the best way to locate them is to run the hand lightly down the trunk until the soft retreat can be felt. The tunnel is constructed with a strong and resistant silk into which are incorporated pieces of material from the bark so that the finished retreat merges completely with the surface. The whole retreat is usually constructed on the surface of the bark but at times a crevice may be used so that only the lid is visible on the surface.

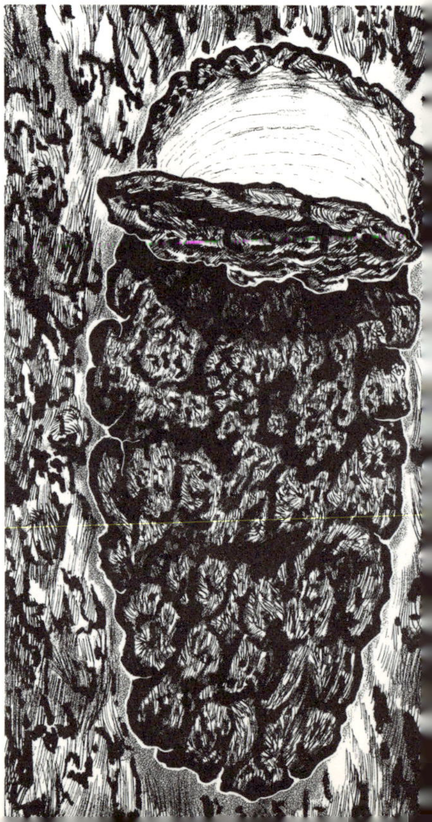

Fig. 37 Nest of the tree trapdoor spider *Migas kochi*.

The True Spiders – Araneomorpha
Hunting Spiders

21. Riverbed Wolf spider. Family Lycosidae. Ahuriri River, Otago. Length 5 mm.

22. Riverbed Wolf spider carrying an egg-sac. Family Lycosidae. Orari River, Canterbury. Length 5 mm. This species and 21 are typical of a number of as yet undescribed greyish wolf spiders which live among the stones of riverbeds.

23. The Seashore Wolf spider. Male. Family Lycosidae. Chrystalls Beach, Otago. Length 7 mm. This and other similar pale species inhabit the sand beaches and nearby dunes along the coast.

24. *Pardosa goyeni*. Male. Family Lycosidae. Omakau, Otago. Length 8 mm.

25. *Lycosa hilaris* with egg-sac. Family Lycosidae. Dunedin. Length 6 mm.

26. *Lycosa hilaris* with spiderlings emerging from the egg-sac. Family Lycosidae. Dunedin

27. *Lycosa hilaris* with the spiderlings in position and the egg-sac abandoned. Family Lycosidae. Dunedin.

Fig. 38 *Migas saxatilis* from Banks Peninsula. Male above, female below. This species lives on the cliff-faces near the sea where it constructs its nest in cracks in the rock.

It is then not difficult to bridge the gap to those species which have forsaken a protected forest life to live on exposed cliff faces or even in burrows in the open ground in the same way as Ctenizids. Many of these non-forest species live along the coast near the sea, and *Migas distinctus* thrives in both habitats, but other species such as *Migas linburnensis*, found in treeless areas of Central Otago, have taken to burrowing in the ground and live in the same way and in close association with species of typical trapdoor spiders. Unlike the true trapdoor spiders, however, mating takes place during the summer months, for this is when the males become mature. The single egg-sac constructed and attached to the inside wall of the mother's nest may contain, in some species, as few as twenty eggs but in most from one to two hundred. When the spiderlings emerge, they show little desire to explore their new world but settle down close to the home of their parent.

CHAPTER FIVE

The True Spiders

The term 'true spiders' is used generally to include all those spiders which move their fangs transversely rather than up and down, and so in New Zealand it means all spiders other than the trapdoor spiders and their relatives. There is another generalization which covers all the true spiders apart from a very few exceptions and this is that the true spiders do not have more than one pair of book lungs. Although the exceptions number only some seven species found in widely separated parts of the world, it does happen that two of them are found in New Zealand.

It is thought that these so-called Hypochilomorph spiders represent the last survivors of the kinds of spiders which flourished hundreds of millions of years ago and to the student of spiders their discovery creates the same interest which the Tuatara raises among the specialist in reptiles. There was little thought some fifteen years ago when one of us was checking through spiders collected by Mr J.H. Sorensen from the bush near Orepuki in Southland that some of these spiders might live in New Zealand. The three species which were known at that time, one each from Tasmania, China and the United States of America were all very conspicuous spiders with long and slender legs. In addition they spun large webs to trap their prey and so could hardly have been overlooked. The spiders which Mr Sorensen had collected were reddish brown with a black band on the abdomen and at first glance seemed to be wolf spiders. However an examination of the position of the eyes showed that they could not be, but it was not until they had been turned over on their backs that it became obvious that they had four book lungs and so might be associated with the rare Hypochilomorph spiders.

It was not very long before we were on our way to Orepuki so that we could study these animals in their natural home. Then it became clear that they have forsaken the use of a snare and catch their prey by hunting. This explains why they look so much like the typical hunting spiders.

Fig. 39 *Gradungula sorenseni* Family Gradungulidae. This four-lunged true spider which hunts its prey on the forest floor is known only from Australia and New Zealand.

Eventually this spider was named in honour of Mr Sorensen as *Gradungula sorenseni* (Figure 39) and, because it was so different from all the other possible relatives, a new family Gradungulidae was established for it. However, Sorensen's spider was not alone in this family because at about that time Dr T.E. Woodward from the University of Queensland found another spider living in the forest near Brisbane in Australia. This was so similar that a very detailed examination under a microscope was needed to make sure that it was in any way different. The fact that these two spiders, which had been separated by a wide expanse of ocean for many millions of years, should still be so alike caused no surprise because it is often found that these ancient survivors of earlier geological periods evolve very slowly.

The most surprising chapter to the story of these spiders was begun with the discovery a few years ago by some of the enthusiastic speleologists who are systematically exploring our caves that a still further species lives deep in the darkness of some caves in the South Island. This spider, which is still unnamed, has remained a hunter but has developed the long and spindly legs so typical of the overseas Hypochilomorphs. Although the body of the cave spider is a moderate inch in length, its legs as they move may span a full five inches, making this the largest spider known in New Zealand, (Figure 40).

Sorensen's spider hunts on the forest floor at night and during the day shelters under loose bark or fallen logs. It has subsequently been found on Stewart Island and along the entire west coast of the South Island, but it has never been seen in the North Island. At present the Cave Spider cannot be considered common, being known only from two caves in the South Island, but it is hoped that as more caves are explored it will be found to be more widespread. It is not known precisely what is caught for food but they most probably prey on the cave wetas which abound on the walls of the caves they inhabit. The eggs, laid in November, are enclosed in a huge, inch-long, white egg-sac shaped like a pear and hung from the walls by a slender stem, but apart from this fact nothing is known of the life histories of either of our representatives of this fascinating and scientifically important group.

HUNTING SPIDERS

Although many diverse groups of spiders which actively hunt their prey instead of spinning silken snares are described as hunting spiders, in New Zealand the term first brings to mind two particular groups – the Lycosids or Wolf spiders and the Pisaurids or Nurseryweb spiders.
Both of these families are abundantly represented in most overseas

Fig. 40 *Gradungula* sp. with its egg-sac on the wall of a cave.

countries and this is also true of New Zealand. As might be expected for such a prominent group much has been written of their habits, but surprisingly little is known about the life histories of our own species and in fact very few of them are as yet officially known to science. For this reason we have spent many interesting and productive hours watching these spiders both in their natural habitats and in captivity, trying to gather sufficient information to write this chapter.

LYCOSIDAE

The chances are that the spider you see dashing across the garden path or between the plants on a sunny summer's day is one of the common Wolf spiders, for these are among the few fast-moving spiders which will hunt in the open during the daylight hours. A careful look at specimens of these spiders from different parts of the country shows that, although they are very similar in appearance and habit, they are in fact different species. The initial impression of a rather dull and unattractive brownish or grey animal is dispelled on closer examination. The strong legs of the female are banded and mottled with yellow, green, black and brown. The carapace is attractively patterned with yellow or white hairs relieved by two broad black bands which do not quite meet in the normally bright yellow centre line. Thc short fur-like pile on the abdomen is set off on the front half by a vivid yellow or orange stripe down the centre, accentuated by a cluster of black hairs along the margins. The male is similar in appearance but because the body is smaller it seems longer-legged, while the stripe on the abdomen is often much wider and longer. Wolf spiders are always fast and agile runners, but if one is watched closely you will see that it always runs in short bursts and then pauses, as if to get its breath back, before dashing off again. It is during these pauses that many of the accompanying photographs have been taken. In the matter of their structure and habits the great majority of wolf spiders are fairly uniform; thus an account of the habits of the common striped wolf spider from our garden in Dunedin can be used as a guide to the behaviour of the scores of other species found in New Zealand.

The wolf spiders are true hunting spiders, because, with very few exceptions, of which none is known from New Zealand, they catch their prey 'on the hoof' rather than construct a web and wait for the food to entrap itself. Many references are made to the keen sight of these spiders, but this attribute is probably not fully deserved, particularly if we compare them with those paragons of vision in the spider world, the jumping spiders. The most useful characteristic by which to recognise a wolf spider is the arrangement of the eyes (Figure 41). There are always eight, grouped in three rows. The four front eyes are small and because they are placed on the front face of the carapace which slopes down, their field of vision is to the front. The next pair are large and situated on the side of the head so that their field of vision is slightly to the front as well as to the side, while the large hind pairs are placed well back and are directed to the side. It is thought that the two posterior pairs of eyes are concerned mainly with the detection of movement and it is this ability

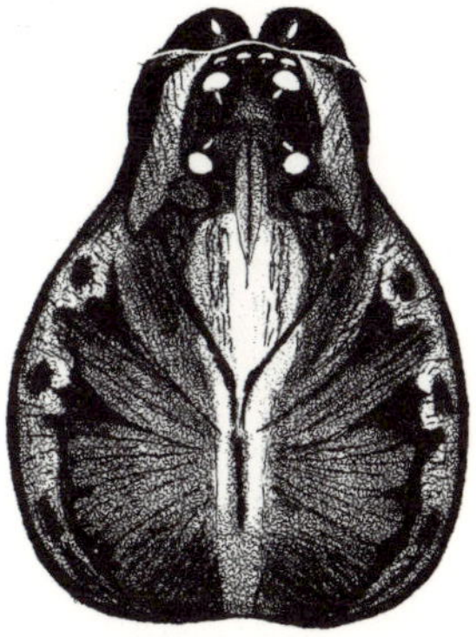

Fig. 41 Carapace of a) Wolf spider b) Nurseryweb spiders, showing the characteristic grouping of the eyes.

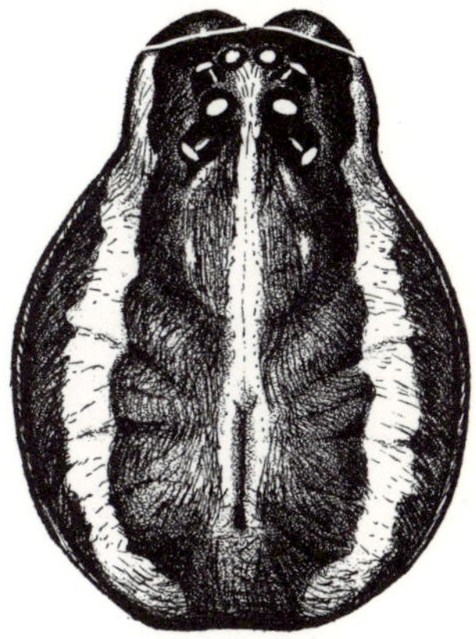

which is the strength of the wolf spiders. The jumping spiders have progressed in another direction by the development of the front pair of eyes whereby they can apparently focus on their prey. The assumption is that they can perceive actual form and also estimate distances.

When watching wolf spiders catch insects, it soon becomes evident that it is not only the eyes which are used to detect the presence of prey, for the spider will pause and wait as an insect approaches, even when it could not possibty have seen any movement at all. This suggests that it is also extremely sensitive to vibrations. In fact, it is only when an insect is a few inches distant that the movement is detected and then the spider exhibits the sudden speed and certainty of action which earn for it the name of 'wolf'. Unlike the jumping spiders, it does not turn its carapace around to face its prey, but closes the gap with a lightning dash so that the next thing seen is the spider clasping the insect with its two front pairs of legs while the fangs are plunged in to inject the paralysing poison.

After a few frantic jerks, the insect reacts with only an occasional twitch and the spider's legs are then withdrawn. Feeding begins while the prey is still transfixed by the fangs. Depending on the size of the meal, it may take from five minutes to over half an hour to reduce the insect to the final soggy little ball of indigestible crushed cuticle and wings which is then dropped and left behind as the spider continues on its quest for further food. With many of the web spiders, it is possible to collect the remains of their webs and so find out easily what they eat, but the wolf spider is a wanderer and this method cannot be used. In captivity, we found that they would tackle nearly everything given to them except, unfortunately, the slaters which happen to be the most abundant source of food in our garden.

Although the striped wolf spider is sometimes found sheltering in a silk-lined tunnel, it does not construct a permanent tunnel home, but normally shelters in cracks and crevices or under stones. The more elaborate silk-lined tunnels seem to be constructed for special purposes such as moulting, or for the period when the spiderlings are due to emerge from the egg-sac.

The methods used by the wolf spiders to care for the egg-sac and then the spiderlings when they emerge are well known and are generally peculiar to this family. The egg-sac is carried along behind the mother attached to the spinnerets; while very few other spiders have this habit one of the exceptions is found in New Zealand. This is a minute theridiid spider which uses this method to shift the egg-sac from one place to another, but most of the time its sac is attached near the web.

The wolf spider's egg-sac is always constructed at night. Early in the evening, the now portly spider secretes herself away and begins to lay down a sheet of silk. This sheet is about one and a half inches in diameter, roughly circular in outline and is held firmly to the ground by numerous attachment threads around the margin. After the sheet is thick enough to support them, the eggs are laid in the centre by the spider as she stands on tiptoes with the abdomen bent downwards. The eggs ooze out in a glutinous mass from the opening on the underside of the abdomen and settle as a small pink ball in which all the eggs adhere together. Immediately after the eggs are laid, the mother begins to lay down the upper sheet which will seal the eggs within the cocoon. At first the half-made sac looks like a lightly poached egg in miniature, but as the silk is laid it becomes a white bi-convex disc. The placing of this upper layer of silk is by no means haphazard, but only by careful observation can it be seen how she achieves such a perfect result. After the eggs have been deposited, she stands over the mound on the tips of her legs and then by lowering her abdomen touches the eggs with her spinnerets. She then backs down the dome of eggs, keeping the spinnerets on the sloping surface. It would seem that as soon as the spinnerets cease to move down and reach the level surface of the margin of the lower sheet, a new behaviour pattern is triggered, because now she begins to exude silk. The silk is laid on with an upward, stroking movement by the manipulation of the abdomen. After the first strand is laid, she then raises her abdomen, bends it to the left or to the right and lays another strand. At each position she lays down three strands which overlap each other. After the three movements have been completed, she moves further up towards the centre of the sheet and goes through these movements again. This performance is repeated until the spinnerets reach the top of the eggmound and then she stops, turns around to face in another direction and edges back to the margin of the sheet where she repeats the procedure. It takes much longer to describe her activities than it does for her to perform them. Each set of actions from the margin of the sac to the apex only takes some forty or fifty seconds, but as each layer is very thin the whole process is repeated again and again so that two or more hours may pass before the egg-sac is completed. The stage of completion is reached for the spider when the slope down the eggs is reduced by the bridging of the silk she has laid down; this is marked by another abrupt change in her activities.

It was this next stage puzzled us most for here she now had a disc firmly attached to the ground which had yet to be transformed into the familiar free, globular egg-sac ready to be attached to her spinnerets. Actually her methods prove very straight forward. She turns round astride the sac so that her fangs are over the margin and with them begins to tear at the attachment holding the sac to the ground. As some of the threads become free she slips her palps under the margin and by raising herself up on her legs tugs away to free it further. Sometimes the two front pairs of legs are also slipped under while she raises herself on the hind two pairs. As one portion becomes free, she moves to another sector and quickly pulls htis free. When most of the threads have been detached, she grasps the whole sac with the front two pairs of legs. With a sudden tug, it parts from the ground and she holds it firmly with her legs beneath the body. At this stage it becomes clear how the egg-sac attains its ball-like form, because the silken cover clings to the eggs as she revolves the sac with her legs while a little more silk is laid. It now also becomes obvious why the Lycosid egg-sac has a more or less conspicuous ridge around it, as this is the seam where the upper and lower sheet are joined. This ridge remains throughout the life of the sac; it is usually here that the sac bursts to let out the spiderlings for the next stage of their life history. All that remains to be done now is for the mother to lower her abdomen to touch the egg-sac with her spinnerets and after a few dabs of silk she moves off with her egg-sac dangling behind. Once the egg-sac is attached the spider carries on life as before, concentrating her efforts on the continual task of catching food.

To watch the Lycosids dashing about with the egg-sac bouncing behind makes one wonder just how the eggs and the spiderlings could possibly survive this treatment for the five weeks still to come. However, the silken cover, although thin, is resilient and because the mother has now acquired a single-minded need to have something attached to her spinnerets the sac will rarely be lost. This urge is very strong for a week or so after the egg-sac is constructed, so much so that if it is removed and given back a day or so later she will grasp it in her legs as soon as she touches it with such a sense of urgency that one might think of her saying 'thank goodness my long-lost egg-sac has been found'. However a few simple experiments soon show that little more is involved than an urge to carry a roundish object. She will in fact accept another spider's egg-sac including those of quite different spiders and even a ball of cotton-wool or cork will be accepted as long as it is round and not too heavy. In the early stages of development of the eggs this drive is strong and she will accept her sac even after two days absence. As time goes on the urge becomes weaker and if the egg-sac is then removed, it is quite likely that she will abandon the cares of motherhood completely.

The actual time which elapses in the case of the striped wolf spider between the laying of the eggs and the emergence of the spiderlings is five weeks. You will notice that we do not say that it takes five weeks for the eggs to hatch because in fact they hatch in just under four weeks. During the last week the spiderlings are completely helpless and remain quiescent inside the sac, but at the end of this period they undergo their first moult and are then ready to emerge. It is not possible to watch the development of the eggs through the opaque silken covering, but if the sac is taken from the spider a small hole can be cut through the silk to see what is happening. When the sac is given back to the mother, she will inspect it carefully and seal the opening before placing it back on to her spinnerets.

Just before the spiderlings begin to emerge the spider will often spin a silken retreat in which she remains until the egg-sac is empty. By this time the once gleaming white egg-sac has become a dirty grey and the silken cover is stretched tight. The sac bursts in the early evening and always at the seam though it has been reported for species in other countries that the mother actually bites it open. This we never saw and, although the mother could at times be seen pecking away with her fangs, the sacs always seemed to burst open themselves, probably due to the pressure exerted by the spiderlings now moving about inside.

At this stage the spiderlings are colourless except for the conspicuous black pigment around the eyes. Their emergence is slow, so that it would be two or three hours before the first few moved out and early morning before the sac was completely empty. The spiderlings counted emerging from a number of egg-sacs of the striped wolf spider ranged from eighty to one hundred and thirty five, and so in round figures the mother has to cope with a hundred or more of these little spiders over the next stage of their life history during which time they cling tightly to her body. The initial task faced by the spiderlings after they have managed to clamber out of the egg-sac is to make their way safely to the mother herself. Fortunately for them, the spider is usually quiescent at this time and those which get off on the wrong side and reach the ground usually touch a leg as they wander about and use this as a guideline to reach her back. Because there is rarely enough room for all of the spiderlings to cling, a spider with a recently emerged brood looks completely top-heavy with the spiderlings two or three deep. However it is not long before the load is reduced to more acceptable proportions because the mother spider shows little or no maternal solicitude. The spiderlings are ignored unless they stray forward over her eyes when, with a flick of her front legs they are unceremoniously swept out of the way. The fact that some of her passengers might occasionally lose their footing and fall by the wayside seems of no concern; unless they are lucky enough to grasp a leg and crawl up again, they are left behind. The spiderlings do not feed at this stage and so do not attack each other, but in a week or so, when they are ready to undergo a further moult and then enter their first feeding stage they rapidly disperse to lead their own individual lives. If the mother spider does not help or feed the spiderlings during this time, it may be wondered just what advantages this chapter in their life history brings but it is probable that the quick reactions of the mother to danger during this vulnerable period adds considerably to their chances of survival. In addition the wolf spiders have a critical humidity requirement which causes them to react to any detrimental change, thereby helping the young avoid dehydration at a stage before their own built-in mechanism becomes functional.

The striped wolf spider is at home in the open country, the grasslands and tussock country and is one of the few which has become adapted to life in the changed habitat of our pastures and gardens. Another wolf

Fig. 42 Stages in the life history of the wolf spider a) After the eggs are laid the top cover of silk is spun. b) The completed egg-sac is tugged free. c) The egg-sac is clutched with the front legs and palp and rolled until it takes up its proper form. d) A few dabs of silk fixes it firmly to the spinnerets. e) The mother spider is now free to run about. f) The egg-sac splits and the spiderlings slowly begin to emerge. g) The spiderlings crawl out. h) They reach the abdomen of the mother and cling in an untidy heap. i) Once the egg-sac is empty it is left behind.

spider which is occasionally found in the garden, hidden away in damp places under boards or at the base of clumps of grass is the Brown Wolf spider (*Allotrochosina schauinslandi*) (Figure 43). This is primarily a forest species, and in fact our only forest dweller, which is dependent on continuous high humidity to exist, dying within a short time if kept in a dry container. During the daytime it can be found hidden under rotting logs or among the leaf litter on the forest floor, but at night it comes out to hunt. During courtship, the male raises his front pair of legs in the air, and as he approaches the female, bends them into a U shape and at intervals vibrates them so fast it is difficult to see the legs moving at all. If the female is not willing to accept his advances she raises her front legs, waggles them in unison and then the male will usually run away.

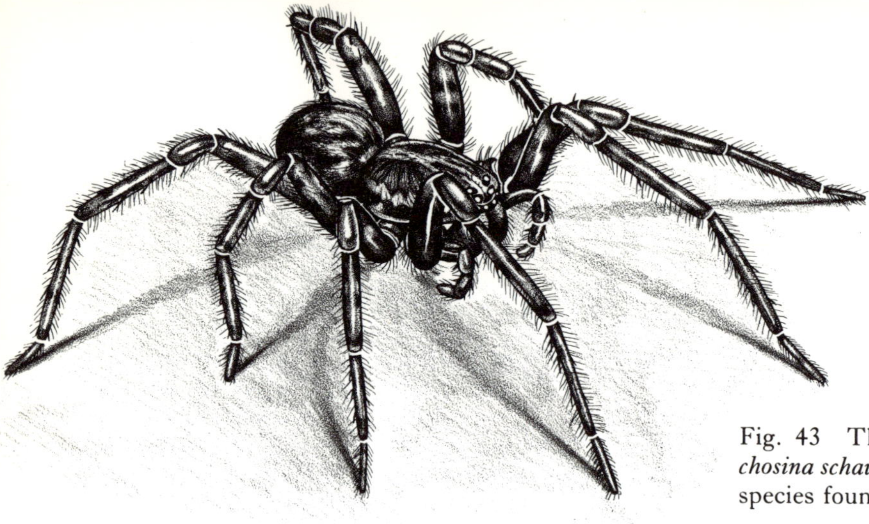

Fig. 43 The Brown Wolf spider *Allotrochosina schauinslandi* is the only New Zealand species found in the forest.

This movement of the legs seems to be the threat or defiance signal which may be used by either the male or female when they meet at inopportune times and immediately causes them to move on their way. The adult spiders never seem to attack each other although an adult will attack a younger spider. If the male is accepted he creeps over the female's head and reaches down over the side of her abdomen to insert his pedipalp first one side and then on the other. The whole action is extremely rapid and is over in less than a minute. The egg-sac has a bluish tinge, is smaller than that of the striped wolf spider, and is carried under the abdomen rather than dragged behind.

A considerable range of species is found in the South Island riverbeds. The slatey grey *Geolycosa arenivaga* can be seen running over the hot rocks in the heat of the sun capturing the small insects sheltering between stones. In addition there is a larger nocturnal species *Pardosa goyeni*, fawn in colour with a vivid yellow stripe on the anterior portion of the abdomen in the male, but duller in the female. Neither of these spiders seems to make a burrow, but in September the females of *goyeni* are found in scooped-out depressions under the stones in the riverbed with their egg-sacs. Perhaps this species remains secreted away with its egg-sac until the spiderlings emerge. Another score or more species are found only in riverbeds, but none of these have yet been officially recorded. This is also the present fate of the scores of other species which live on the exposed mountainsides above the bushline and along our beaches. Most of the mountain species are very difficult to see among the stones until they move, but it is on the sandy beaches and also among the sand-dunes that the most extraordinary examples of camouflage are found. The Sand Wolf spiders are all pale cream dotted with darker patches which break up the outline of the legs and body. A scurrying movement which looks like a puff of sand might catch your eye, but when you reach the spot where the movement stopped there is nothing to be seen. The spider will remain motionless with its legs outspread against the sand and thus blend perfectly with the background. None of the spiders which we have mentioned regularly constructs a tunnel retreat, but there is at least one species found in the inland area of Otago, which digs a nest two or three inches deep and lines it with a thin coating of silk. This spider may also build up a rim of silk in which it entangles nearby pieces of grass. The habits of this Turret Wolf spider (*Pardosa bellicosa*) are probably very much the same as those spiders which live in a similar way in other countries. There the spider sits at the opening of the turret and with the approach of an insect dashes out and returns with it to the retreat. When the female spider is carrying her egg-sac she will come regularly to the mouth of the turret and sit with her head down so that the sac pokes out through the opening into the sun. In captivity, all that was needed to make the spider do this was to bring the electric lamp close and within a very short time the sac could be seen pushing aside the few threads which held the opening closed.

PISAURIDAE

The huge gleaming white nursery webs on the tips of low shrubs such as gorse, broom or manuka are familiar to everyone, but while most people know that these webs have been made by a spider how many have actually seen the spider itself? (Figures 45–47). The Nurseryweb spider is nocturnal in its habits, so if you wish to see the architect you will need to come out at night with a torch. There you will find her clinging to the web, having crept out from her daytime hideout to take up her protective stand for the evening hours. She is a magnificent creature whose body seems clothed with the finest velvet. Her legs are brown and the carapace is patterned on each side with three longitudinal bands, first brown, then yellow, and finally black where it reaches the narrow yellow stripe down the centre. The central stripe of the carapace is carried down the middle of the abdomen where it is part of a broad brown band extending back to the spinnerets and broken only with pairs of small white spots. The sides and lower surface of the abdomen are paler but speckled with brown. The spiders always seem huge because, although the body may be only three-quarters of an inch long, the legs span two inches or more. Spiders belonging to this family are found all over the world and because they are usually associated with water are often called Fisher or Swamp spiders but the common New Zealand Nurseryweb spider (*Dolomedes minor*) is one of the exceptions. Although it is as commonly found in swamps as it is in the dry scrubland,

Fig. 44 Nurseryweb of *Dolomedes minor*.

Fig. 45 At night the mother climbs up and guards the nurseryweb.

29. The Water Spider *Dolomedes aquaticus*. Female with egg-sac. Family Pisauridae. Manuherikia Riverbed. Length 23 mm.

30. *Dolomedes aquaticus* at the water's edge. Family Pisauridae. Shingle Creek, Otago.

31. Nurseryweb of *Dolomedes aquaticus*. Oreti Riverbed.

28. The Nurseryweb spider *Dolomedes minor*, Female. Family Pisauridae. Nenthorn, Otago. Length 19 mm.

32–34. The final moult of a male *Dolomedes aquaticus*. Karangarua Riverbed.

36. Undescribed arboreal toxopid. Female. Family Toxopidae. Port Chalmers. Length 2 mm.

37. *Oxyopes* sp. Female. Family Oxyopidae. Waipiata. Length 5 mm.

35. *Cycloctenus* sp. Female. Family Toxopidae. Motueka. Length 8 mm.

38. *Argoctenus* sp. Female. Family Ctenidae. Foxton. Length 7 mm.

39. Arboreal *Clubiona*. Female. Family Clubionidae. Dunedin. Length 9 mm.

40. Terrestrial *Clubiona*. Female. Family Clubionidae. Palmerston. Length 6 mm.

41. *Clubiona* mating. Kaitangata. Length 9 mm.

42. *Miturga* sp. Female with egg-sac. Family Clubionidae. Pohangina. Length 20 mm.

43. *Miturga* sp. Female. Family Clubionidae. Te Wharau, Wairarapa. Length 16 mm.

44. *Chiracanthium stratioticum*. Female guarding egg-sac within retreat. Family Clubionidae. Nelson. Length 9 mm.

Fig. 46. The prey of *Dolomedes minor* is firmly transfixed to the fangs.

Fig. 47 The egg-sac of the Nurseryweb spider is held under the cephalothorax with the fangs and palps.

Fig. 48 The egg-sac is held firmly beneath the cephalothorax.

it does not seem to require free water to thrive and its webs are a common sight along the sides of the road and on the scrub-covered hillsides. We do have another group of these spiders found only near water, usually in the beds of rivers and streams or along the margins of lakes. They were well known during the goldmining days of Central Otago. Large numbers made their homes on the water races the miners constructed for sluicing and consequently were called the Water spiders. P. Goyen, an early Inspector of Schools in Otago, who studied spiders on his visits to schools in his district, honoured this name when he published an account of its habits in 1887 under the new scientific name, *Dolomedes aquaticus*. The habits of this spider seem closer to those overseas species than to the common nurseryweb spider and will be described later.

The nurseryweb spider is found wherever there is low cover, particularly scrub, long grass or sedges but it seems that it does not like full tree cover because it is never seen deep in forest. While it is more common in the lowlands it is also found high up in the mountains where nurseries can often be seen binding together the tips of the tussocks and swaying in every breeze.

Nurseryweb spiders catch their prey in a manner similar to wolf spiders. They seem very sensitive to vibrations and will pause as an insect nears, taking up a stance with the two front pairs of legs stretched forward. Little further action occurs until the insect is very close and often not until it actually touches the spider but the ensuing action is extremely rapid. With a quick flurry of movement, the spider pounces on top of the insect, clutching it tightly under the cephalothorax with all four pairs of legs. The fangs are thrust in at once but as the victim's struggles decrease the legs are withdrawn and it is then held firmly in place by the fangs and palp (Figure 46). As digestive fluid oozes from the mouth, the fangs open and close and the chelicerae themselves move out and then in, crushing the body of the prey. The insect is rapidly crushed into a small moist ball which is turned round and round as fluid is alternately exuded from and sucked into the mouth. Eventually all that is left are the indigestible fragments which are then discarded. According to the size of the catch, anything from ten minutes to half an hour may elapse from the time the insect was captured to the completion of the meal. Surprisingly the spider may then spend another half hour just sitting and grooming itself. It will rub its palps together then pass each one through the angle between the fang and the base of the chelicera. It will do this repeatedly and then go through the procedure again with the front pair of legs. Often the end of this grooming is marked by a sweep of each palp back over its eyes with a movement just like

Fig. 49 The Water spider *(Dolomedes aquaticus)* waits by the side of a stream with the front two pairs of legs resting on the water.

someone brushing the hair from his eyes. When this toilet is completed, the spider once more sallies forth in its search for food.

It is in the late spring or early summer that the female spiders are found, each holding by the fangs under the cephalothorax a huge egg-sac which by its very size forces them to move about on tiptoes (Figure 47). The mother spider carries this egg-sac for at least five weeks before the spiderlings emerge. Because she seems to have committed the full use of her fangs to this new task, it may be thought that she must now undergo an enforced period of fasting. On the contrary, any insect which comes her way suffers the same fate as before. It is clasped with the legs at the same time as the sac is released from the fangs and pushed back under the body by the tips of the palps thus leaving the fangs free for action. Sometimes the spider will even release the egg-sac entirely but it will still keep it close so that at least one leg is touching it lightly. At the slightest disturbance the sac is scooped up, clasped firmly beneath her body and she dashes nimbly away carrying both sac and prey without any noticeable reduction in speed or agility. Towards the end of five weeks, the egg-sac loses its pristine whiteness and begins to turn grey. The mother can be seen pecking away with her fangs at the surface covering although not actually breaking through to the spiderlings within. The spiderlings have hatched a week or more before but now they are due for their first moult and after this is completed they will need to leave the sac. How the mother spider can judge the time when the spiderlings are ready to hatch is a puzzle. Does she detect movement in the sac or has she some biological clock which signals to her that the requisite time from the laying of the eggs has now elapsed? Whatever may be the mechanism at this time, her whole behaviour changes. She ceases to catch food, climbs up to the top of some adjacent shrub or clump of grass, firmly attaches the egg-sac, and spins the bulky nurseryweb around it (Figure 44). This is always done at night and although the egg-sac is now disposed of and protected within the web she still does not desert her young, but clings faithfully to the side of the nest. She will remain with the nest for another week, creeping down to hide at the base of the shrub during daylight hours but resuming her vigil as darkness comes (Figure 45). The spiderlings begin to emerge from the egg-sac shortly after the nursery is completed and move freely within, leaving threads of silk as they go, fully protected from the vagaries of the weather and any hungry predators who might be near. They remain within the nurseryweb for about a week, after which the silk towards the tip becomes broken, enabling them to emerge. The growing tips of the shrubs sometimes pull the web apart. It is said that the mother spider herself opens it up but we have found that the spiderlings will still escape when the mother is removed. Even after the spiderlings have left the nest, they show no eagerness to move away and for another week they may be seen clinging nearby, their draglines merging with the silk of the nursery so that for a while this seems to grow even larger. Slowly they depart to lead their own lives until, at the end of ten days, all that remains is the old egg-sac, some minute cast skins and the tattered remnant of the once gleaming nest.

THE WATER SPIDERS

The water spiders are slightly bigger than the nurseryweb spiders but their colouring is a little more subdued and matches more closely the stones among which they are found. The legs of the common Otago species, *Dolomedes aquaticus* Goyen, are fawn but the carapace is strikingly patterned with a narrow yellowish band down each margin while the inner area is blackish brown. The conspicuous yellow band down the middle of the carapace of the nurseryweb spider is represented in this spider by a much narrower band which can barely be seen. The abdomen has a broad greyish band down the upper surface and the prominent yellow streak of the nurseryweb spider is missing. The male is much paler and being more slender in body has a lanky appearance and seems even more agile than the female.

During the daylight hours, these spiders shelter under the stones near the water's edge and so are rarely seen despite the fact that wherever the conditions are right they exist in vast numbers. They feed at night and a cautious visit to the water's edge with a torch will show them clinging motionless to the surface of the stones with their two front pairs of legs stretched out on the surface of the water, where the numerous fine hairs enable the limbs to rest without breaking the surface film (Figure 49). An incautious step or a stumble among the stones and the spider is gone. It is not to be seen on the surface of the water – it has run down the side of the rock below the surface and there it can be seen gleaming in the light of the torch from the hundreds of small bubbles of air trapped among its hairs (Figure 50). The spiders are able to stay beneath the surface for surprisingly long periods of time. Some we watched were still submerged after half an hour and showed no signs of surfacing, breathing the air held by the bubbles around the openings to the tracheae and book lungs. But it does not seem that they normally catch their prey under the water, although they would no doubt grasp anything unwary enough to come within their reach. If the spider is

Fig. 50 Clothed in a gleaming coat of bubbles the Water spider clings to the rock face beneath the water.

given a slight push while under the water, thus losing its grip, it will come flashing to the surface, often on its back. The water immediately runs off and the spider, then high and dry, rests on the surface (Figure 51). If it lands on its back then with a quick flip it turns the right way up. While we have never seen a water spider catch anything under the water, it is certain to behave in its normal fashion by clasping the prey with all its legs. As soon as it lost its footing, it would shoot to the surface and perhaps this is what it does do. When we read accounts of the behaviour of close relatives of the water spider from other countries, we found that these spiders are said to hunt on the surface of the water. They rest with the front legs on the surface until the ripples caused by some struggling insect are detected and then dart forth over the surface to capture their prey. At the same time they are supposed to let out a dragline by which they may pull themselves back to the same spot on shore. The behaviour of the spiders we watched was not quite so spectacular. They certainly rest at the water's edge with the front two pairs of legs on the water surface but show not the slightest response to anything on the water until it actually floats so close it touches one of the legs. Then the victim is immediately scooped up and if during this movement the spider slips into the water it extends its legs and floats downstream.

As soon as it touches dry land, it clambers out of the water to take up its stance once again with the front legs outstretched. The food is eaten in the same way as we described for the nurseryweb spider. These spiders are also to be seen floating on the water with the legs widely outstretched drifting with the current. This seems to be another way by which prey is captured because contact with a floating insect during this time inevitably results in its immediate seizure. Even with all of its legs tucked under its body, the spider still manages to float high in the water. In a short time, the legs are again extended while the spider floats with the insect transfixed to its fangs. Although some of the spiders then float passively until they reach the shore where the meal begins, others take a more positive action. By using the second and third pairs of legs as sculls they can move over the water at a tremendous speed, and, in fact, if disturbed, will raise their bodies from the water and run over the surface as if they were still on dry land. The spiders will often use this method of progression if they drift into rough water and may even be seen leaping and jumping as they go. We were curious to see whether they could actually eat their prey on the surface of the water because we reasoned that the digestive fluids would be dispersed through the water before they could break down the body tissues and

Fig. 51 With legs outstretched the Water spider floats gently down the stream.

Fig. 52 The meal has been completed and the careful toilet begins.

even if they acted the meal would be a very watery one. However, in all of the instances we observed, the spider eventually came back to shore before beginning its meal.

After meeting with no success in our attempts to observe these spiders courting and mating in their homes among the stones of the river-bed, we decided to bring them together in a special cage we built for them. Alas, our first attempts were sadly unsuccessful because as soon as the male was introduced the female pounced on top of him and he immediately became her next meal. We tried feeding the female to her full capacity with flies before we brought in the male, but the same thing happened again. Success was attained only once and then at a time when our supply of males was becoming so depleted that we despaired of ever succeeding. This aggressive female behaviour, in spite of the statements in many popular natural history books, is by no means usual and as this could not happen so consistently in nature we knew that there must be some factor which we had overlooked. Even with continued observations, the reason for our lack of success is still obscure. We even considered that like the English nurseryweb spider *Pisaura mirabilis* it might be necessary for the male to bring an offering because with this species the male first catches an insect and after carefully wrapping it up in silk carries it in his chelicerae as he approaches the female. By raising himself up on the tips of his hind legs and balancing on the end of his abdomen he offers this carefully wrapped bundle to the female who, if she is receptive, takes it in her chelicerae and proceeds to eat it as the mating takes place. Unfortunately, our male spiders did not seem in any way interested in this technique. As the sole successful mating was with a newly moulted female, it is probable that these spiders mate immediately after their final moult and subsequently regard any approaching male as simply another meal.

In this instance, the successful mating we did manage to observe was preceded by a very short courtship. The male slowly approached the female as he alternately waggled the front pair of legs up and down. She remained still as he neared and touched her gingerly after which he crept directly over her carapace and bending over her abdomen inserted one palp. Within a few seconds he moved to the other side and inserted the second palp before he quickly withdrew and retreated. It still remains to be seen whether the behaviour we observed is in fact typical of these spiders.

The construction of the egg-sac by our species has never been observed but it will probably be similar to that described by Dr Pierre Bonnet for related species in Europe. With these spiders, when the female is ready to construct the sac she first spins a few threads from which she can hang down. A thick platform of silk is then woven as her body circles around until eventually a small dome is completed. Suspended by silken threads this hangs upside down at an angle. She then lays her eggs into this cup by pressing her abdomen beneath it and quickly

Fig. 53 In spite of her huge egg-sac the Water spider moves with agility among the stones.

Fig. 54 During feeding the egg-sac is pushed back from the mouth.

secures them in position with a swathe of silk. Further silk is added until the eggs are completely enclosed and the sac begins to take its final form. The threads which hold the sac are cut so that she is able to hang from her hind legs with the sac held firmly with her legs and palps. This is revolved while the final covering of silk is added, and then with the egg-sac firmly held in the fangs she can move about freely for the five weeks which will pass before the spiderlings emerge (Figure 53). She still can feed in the same way as the nurseryweb spider and in fact can even move over the water as freely as she did before.

We have always been curious to know whether the water spider constructed a nurseryweb in the same way as her relatives but surprisingly no one seemed to know. It is true that there are always nurserywebs near the streams and rivers and some of them even on low shrubs growing in the riverbeds, but those we examined seemed always to be associated with a nurseryweb spider and not the water spider. It was only after a year's futile searching that the answer came. While investigating a large sheet web between the stones, we found the water spider sitting below it next to her egg-sac and there the problem was solved. Whereas the nurseryweb spider instinctively moves up to the tip of a shrub when her spiderlings are due to appear, it seems the water spider moves down. It attaches the sac in a crevice and weaves the nursery about the egg-sac, bridging the spaces between the stones. Instead of clinging to the outside of this web the mother crouches below with the egg-sac until the spiderlings depart.

CHAPTER SIX

The Ground Hunting Spiders

TOXOPIDAE

In 1940 Professor V.V. Hickman of the University of Tasmania named a new family, Toxopidae, for an unusual little spider he had found living on low shrubs on the summit of Mount Wellington, near Hobart. Apart from a number of unusual internal characters, this spider, which he called *Toxops montanus*, had the back row of eyes strongly procurved and the outer eyes of the front row set back from the front margin of the carapace so that they almost touched the posterior median eyes in the same way as in Cycloctenus (Figure 55). This remarkable spider remained in solitary splendour as the sole member of its family until some twenty-four years later when another twenty-four species of New Zealand spiders which had their eyes arranged in much the same way were placed into this family. Second thoughts on the true relationships of these New Zealand spiders, however, suggest that some of them are not toxopids at all, but nevertheless for the present we will mention all of these below. Most of our native species have been placed in the genus *Toxopsiella*, which consists of two quite different groups. *Toxopsiella minuta* is one of a group of moderately large spiders ranging from five to six millimetres in length which live as vagrants on the forest floor (Figure 56). Apart from the striking eye arrangement, these mottled brown spiders look like many of the agelenids with which they live side by side. The second group of seven species, of which *Toxopsiella otagoa* is typical, are all minute and never much more than two millimetres long (Figure 57). Unlike the larger species, these spiders are to be found on low shrubs and ferns where they can be seen dashing about, capturing the small flies and other insects which alight on the foliage. It seems that it is this second group of small species which is most closely related to the Tasmanian *Toxops*. *Laestrygones proxima* (Figure 58) is an agile, pale

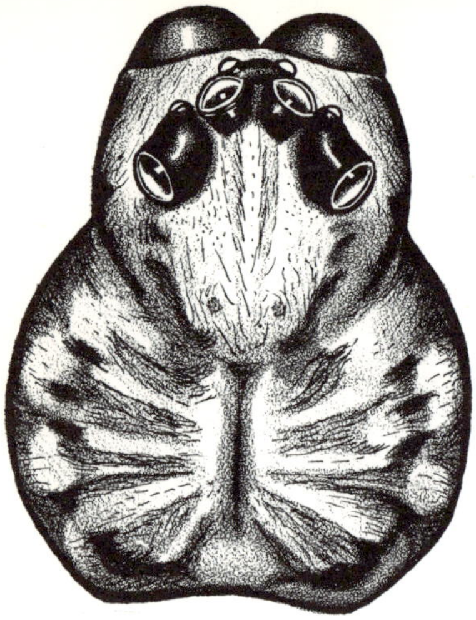

Fig. 55 Carapace of *Cycloctenus*.

Fig. 56. *Toxopsiella minuta* from the West Coast of the South Island lives on the forest floor.

Fig. 57 *Toxopsiella otagoa* hunts on the low foliage in forests.

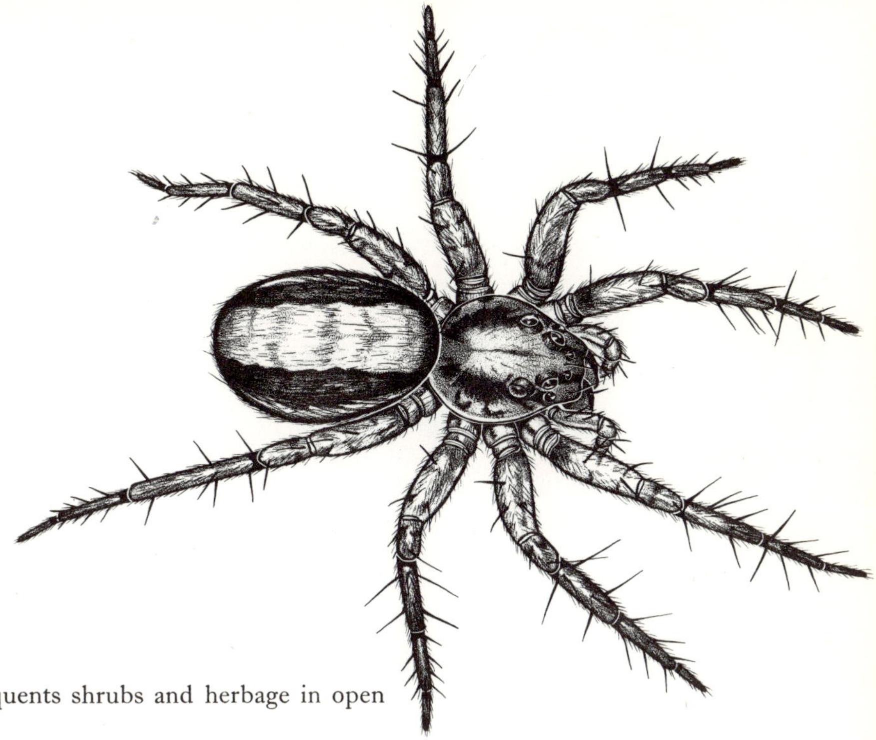

Fig. 58 *Laestrygones proxima* frequents shrubs and herbage in open country or forest margins.

brown spider about the size of *Toxopsiella otagoa* which also has similar habits but, instead of being restricted to the shade of forests, this species frequents shrubs and grass growing in more open country. The fact that it does live in more exposed places probably means that it is able to resist dessication. This species has certainly achieved a very wide distribution, being found all over New Zealand and on the outlying islands, including even the windswept sub-Antarctic islands and the Chatham Islands.

The elongate *Plectophanes* is one of our most odd looking spiders because the front portion of the carapace bearing the eyes projects forward as a long and narrow stalk, well beyond the chelicerae (Figure 59). The possible function of this modification had remained a puzzle for many years because, as an identical feature of both the male and female, it would not be likely to play any part in the mating behaviour in the way

Fig. 59 *Plectophanes* with its elongate body is well fitted for life in narrow burrows in twigs.

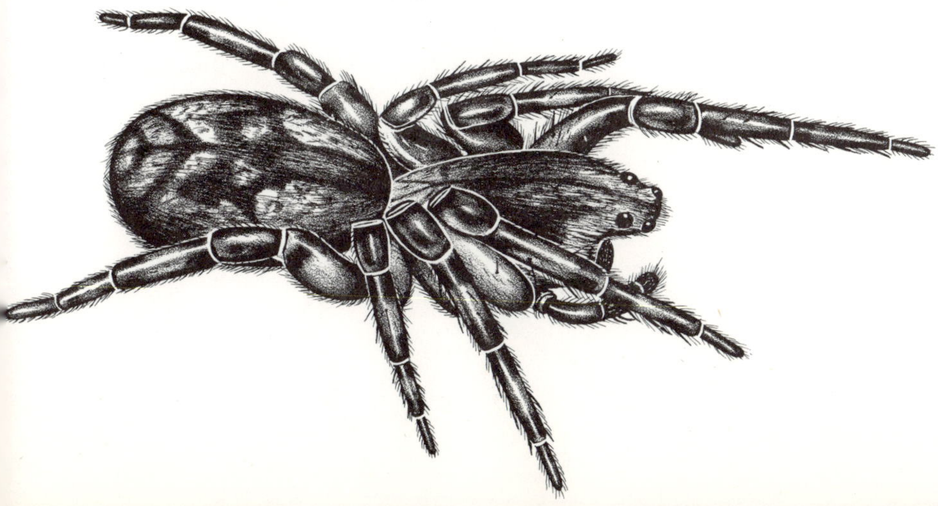

somewhat similar modifications do in other spiders where unusual structures are found only in the males. It was not until C.L. Wilton, during the study of the spiders living on his farm in the Wairarapa, discovered that these spiders always lived in the burrows in twigs and branches of trees vacated by wood-boring insects that direct observations could be made (Figure 60). He found that the spiders took up a position at the mouth of the burrow during the daytime with the projection bearing the eyes peeping out from the hole like a periscope. Unlike most of the tunnel dwellers, these spiders use sight rather than vibrations to detect their prey so that as an insect comes within range there is a flurry of movement as the spider dashes out to capture its victim and return to the protection of its retreat. Before its habits became known, *Plectophanes* was considered to be one of the rarest of spiders but now we know that there are many different species, some of which will be living in any patch of bush in the country. They are not difficult to find, because, although they do not line their retreats with a thick layer of silk or run triplines from the mouth, there is always a little silk near the opening so that the occupied burrows may be recognised.

The final genus placed in this family is *Cycloctenus* (Figure 61). These are most conspicuous spiders but only when they are removed from their preferred hunting grounds on fallen logs and the trunks of trees where

Fig. 61 The mottled colour pattern of *Cycloctenus* makes it difficult to see on the tree-trunks it frequents.

Fig. 60 During the day *Plectophanes* takes up a position enabling it to peer out from the burrow.

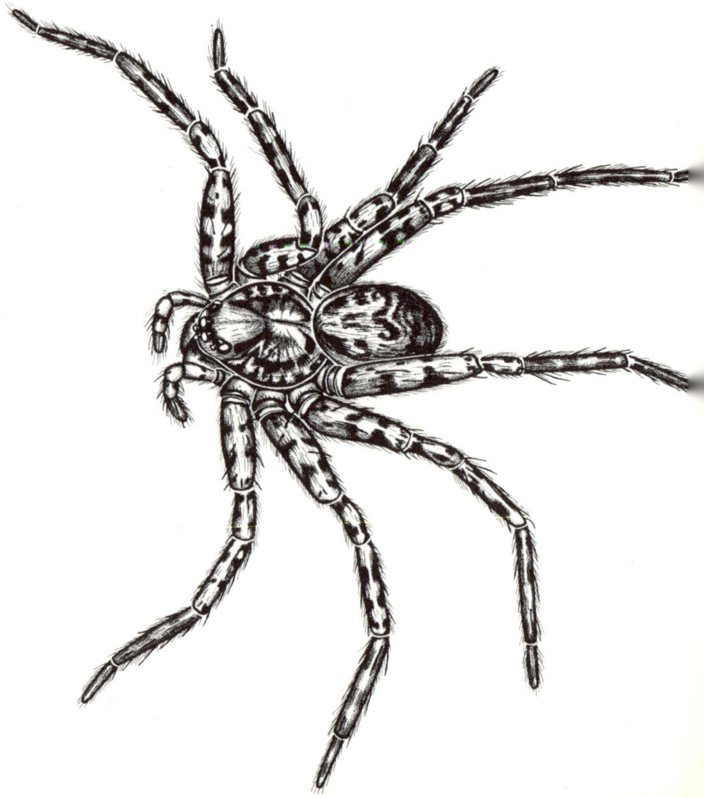

their mottled brown and grey colouring allows them to merge inconspicuously into the background. They are all laterigrade spiders with the legs more or less directed out to the sides and are also somewhat flattened so that they are able to move about rapidly in all directions while they are pressed close to the surface they are on. Catching specimens can be difficult if they happen to be on the trunk of a tree or on a log which is turned over, because before the capture is effected they will have sidled round to the other side. The egg-sacs are stuck on to surfaces wherever they may be and are not always very easy to see as the structure is little more than a low mound from one-half to three-quarters of an inch wide, clothed with a closely woven and smooth silken sheet often covered with pieces of debris.

Placing these spiders in the family Toxopidae has by no means settled the question of their true relationships, although the eye pattern fits in well with the other spiders described above. Originally these were thought to be ctenids, as the generic name suggests, but subsequently it was suggested that perhaps they might be rather unusual wolf spiders and should be placed with the lycosids. The placing of these spiders with the toxopids inspired a German specialist Homann, who was interested in the detailed structure of the eyes of spiders, to examine some specimens sent to him and he concluded that *Cycloctenus* belonged to a predominently tropical family called Telemidae. It must be admitted that this scientific application of the game of musical chairs can be rather confusing to anyone who is not in a position to follow all of the ramifications but eventually order will prevail. In the meantime, it must be understood that all this action indicates the unusual and puzzling nature of many spiders in this country.

ZODARIIDAE

The zodariids are spiders usually associated with more tropical countries and it is surprising that they are represented in New Zealand at all, even though it must be admitted that they are not abundant. Most of them look like our more common ground dwelling agelenids but they are easily recognised because the anterior pair of spinnerets (that is the pair which is in front when the spider is viewed from below) are by far the largest, and the other two pairs may be so small that they are difficult to see. Australia is one of the headquarters for zodariids where many species of both small and large species of *Storena* occur. Most of these have a black abdomen with conspicuous patches, making them easy to recognise. Very recently, a handsome zodariid, which may be a

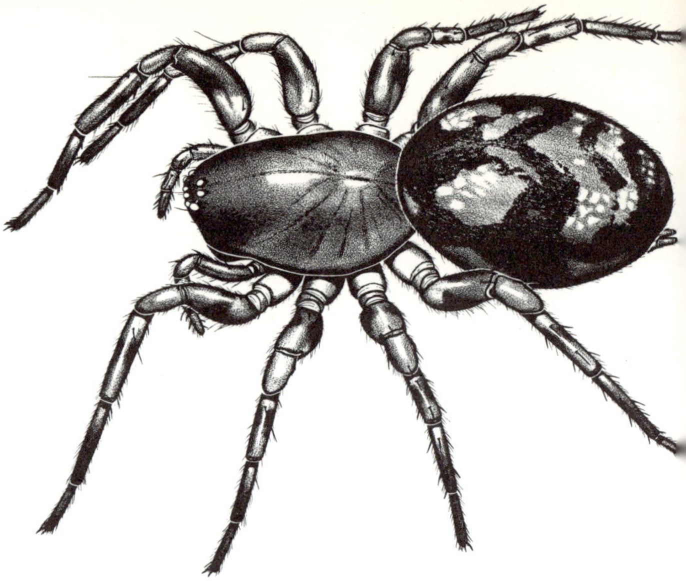

Fig. 62 Numerous species of *Storena* are found in Australia but only this species from North Auckland is known in New Zealand.

species of *Storena*, has been found in the North Auckland district and this may even be a native species (Figure 62). Nothing is known of the habits of this particular species but like its Australian relatives it is probably a terrestrial hunter, or it may construct a tunnel in the ground from which it emerges for occasional forays.

Another zodariid which has been known for a very long time is *Huttonia palpimanoides* which, although only four millimetres long, is a most striking spider with a brick-red carapace and legs and a row of purplish chevrons down the dorsal surface of the abdomen (Figure 63). Unlike most of the zodariids which have a relatively high carapace, in *Huttonia* the carapace is low and, like many other New Zealand spiders, is a most unusual member of its family. It lives a secluded life amongst the debris on the forest floor, coming out only at night to hunt, when its slow and deliberate movements are peculiarly its own.

Fig. 63 *Huttonia palpimanoides* leads a retiring life among fallen leaves on the forest floor.

45. *Amaurobioides maritima*. Male. Family Amaurobiodidae. Matanaka Beach. Length 10 mm.

46. *Scotophaeus pretiosus*. Female. Family Gnaphosidae. Kaituna Valley. Length 9 mm.

47. Banana Spider. *Heteropoda venatoria*. Family Sparassidae. Dunedin, from imported bananas. Length 18 mm.

48. *Isopeda insignis*. Male. Family Sparassidae. Auckland. Length 22 mm.

49. *Sidymella* sp. Female. Family Thomisidae. Catlins. Length 10 mm.

50. *Sidymella* sp. Male. Family Thomisidae. Length 6 mm.

51. *Diaea* sp. Female with parasitic mite. Family Thomisidae. Hunterville. Length 3 mm.

52. *Diaea* sp. Male. Family Thomisidae. Te Wharau. Length 3 mm.

CHAPTER SEVEN

The Lynx, Vagrant and Vagabond Spiders

OXYOPIDAE

Like the zodariids, the Lynx spiders are commonly thought of as spiders of the warmer climes where they may attain considerable size and also be gaily coloured in greens and reds. New Zealand can be considered only a marginal home for the family and our species, as seems fitting, are relatively small and drab. Lynx spiders are specialists for life in the foliage of plants where they hunt actively during the day. The carapace is high and appears large in proportion to the pointed abdomen, while the legs are slender and conspicuously armed with stout spines and bristles which stand out at right angles to the surface. The eyes are always well separated from the chelicerae and arranged in three rows of 2.4.2 (Figure 64).

The straw-coloured *Oxyopes gregarius* is common throughout the North Island where it may be seen running and hopping about on low shrubs and hedges in the bright sunshine of the summer months capturing small insects on the surface of the leaves (Figure 65). Other undescribed species, some of which are silvery white, are found in the South Island but their habits are the same. The egg-sacs, covered with dingy white silk, are attached to twigs frequented by these spiders, and soon after they are constructed the mother may be found clinging to the side as she guards the sac until the spiderlings emerge.

THE CLUBIONIDAE AND GNAPHOSIDAE

Both the clubionids and the gnaphosids are rather drab short-sighted hunters which make extensive use of silken retreats from which most

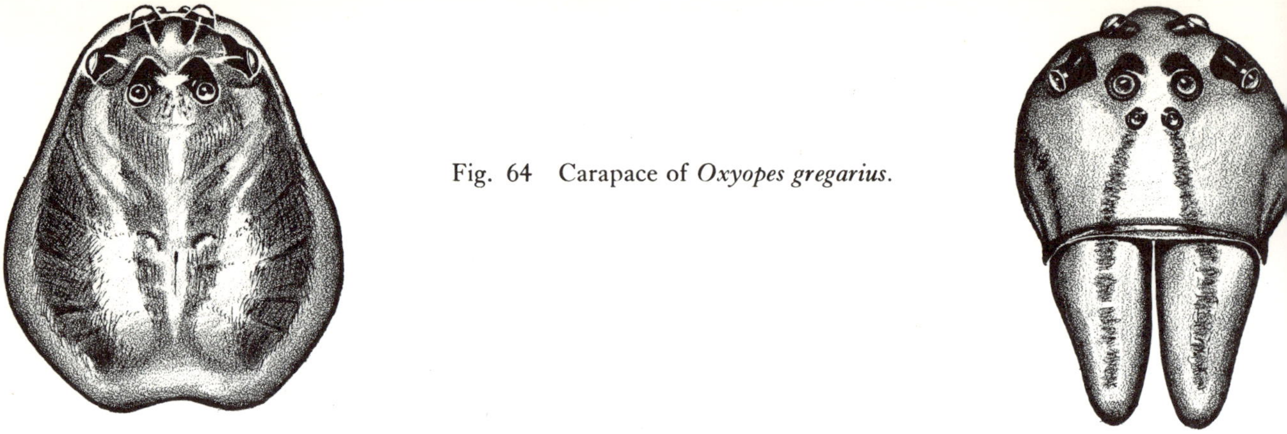

Fig. 64 Carapace of *Oxyopes gregarius.*

emerge at night to capture their prey. Only two claws are present on the tarsi of each leg where there is at least a thick tuft of hair at the tip and often an extensive pad of short hairs along the ventral surface. Spiders of the two families can be confused but many of the gnaphosids are clothed with glistening metallic hairs which are never present in the clubionids and there is a fundamental difference in the form of the spinnerets. Those of the clubionids are conical and usually touch at their base, while those of the gnaphosids are cylindrical and are markedly separate from each other (Figure 66).

CLUBIONIDAE

The three genera of these spiders recorded from New Zealand are fortunately easy to tell apart. The majority, which belong to the world-wide genus Clubiona, are yellowish brown ranging from seven to ten millimetres in length and live in a wide variety of habitats. The native clubionas can be divided into two different groups – those which live mainly on the foliage and those which live on or near the ground. The arboreal species generally has a rather plump and speckled abdomen

Fig. 65 The Lynx spider *(Oxyopes gregarius)* hunts on low plants during the day.

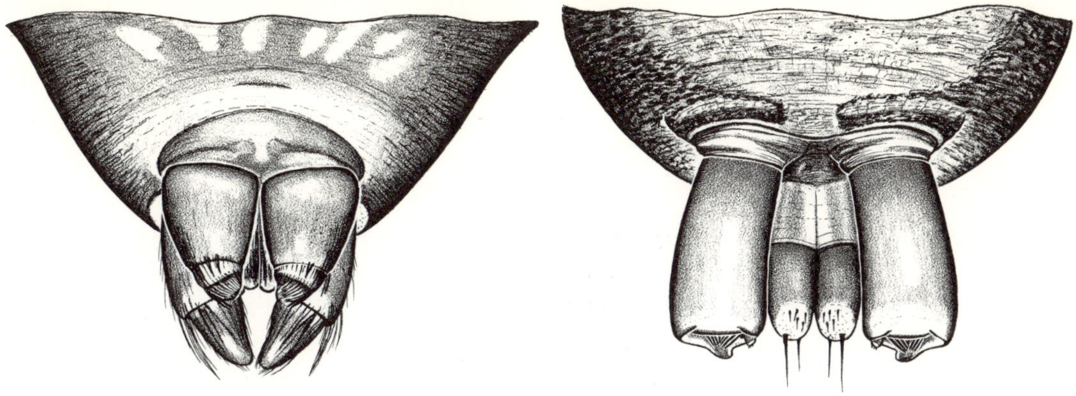

Fig. 66 The simplest way to separate the clubionids from the gnaphosids is by the form and spacing of the spinnerets. a) *Clubiona* b) Gnaphosid.

and is best seen at night when it will be found that the normal steady progression of movement as it hunts is interrupted at times with sudden hops (Figure 67). This hopping is in fact quite different from the jumping of most spiders because instead of moving forward in a nice smooth trajectory, as for instance the jumping spiders, the clubionas seem to lift upwards before they are propelled forward and rarely achieve more than a few body lengths at each hop. Those belonging to the second group of clubionas have a uniformly brown abdomen without spots; an abdomen which is relatively narrower and sometimes pointed behind. These spiders favour terrestrial life and are found under stones or at the bases of clumps of grass and sedge and seem equally at home in dry or swampy areas. Once adult, the spiders apparently live for some time and may be found in any month of the year, but mating and egg-laying take place only in the early and middle summer. Mature males are often found in the retreat with a sub-adult female and so, like a few other spiders, the male may in this way stake his claim to the attentions of his future mate by waiting for her to pass through her final moult. During the actual mating sequence, recognition seems to be based mainly on tactile responses when the male taps the female with his front pairs of legs and then creeps over her as she crouches down, to insert his palps in much the same way and position as we have described for the wolf spiders. The egg-sac which is laid by the female within her retreat is a neat, rounded structure covered with a thin white sheet of silk through which the shape and colour of the eggs can be seen.

The genus, *Chiracanthium* is found throughout the world although the sole New Zealand species *Chiracanthium stratioticum* is found only in this country and Australia. The usual colouration is pale brown,

Fig. 67 The speckled species of *Clubiona* are found mainly on the foliage of low shrubs.

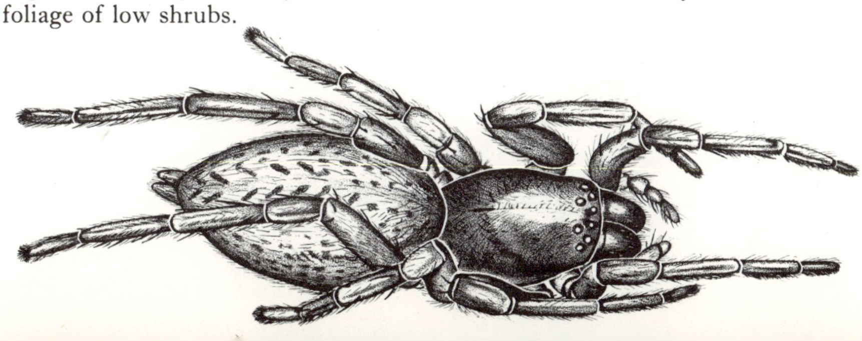

although sometimes specimens are seen with a faint green flush so that this species looks somewhat like the ground-dwelling clubionas but the legs are relatively longer. Its habits, however, are quite different because, although it often lives in clumps of grass as well as in low shrubs, the retreat, rather like a small nurseryweb, is constructed well up off the ground. It has been found that some European species of *Chiracanthium* adopt an unusual posture as the male fertilizes the female, with their ventral surfaces touching but with the two spiders facing in opposite directions. It would be interesting to confirm that this position is also adopted by species from the Southern Hemisphere and perhaps someone in this country might keep a watch on our Chiracanthiums during the early summer when they probably mate. From December to February some of the retreats will be found sealed while the mother spider guards her egg-sac within until the spiderlings are ready to face life on their own.

The bulky spiders placed at present in the genus *Uliodon* are very different from the other two genera (Figure 68). These are all handsome, brownish spiders with the abdomen and sometimes the carapace closely clothed with a thick velvet-like plush of short hairs, and live exclusively on the ground. In spite of their abundance and eye-catching size, they have never attracted any specialist attention and little is known of their habits. In fact they should not really be called *Uliodon* as they are more closely related to the Australian spiders of the genus *Miturga*. The majority, of which there are a score or more unnamed species, are restricted to the forests although they will also thrive where the forest is cleared as long as a few fallen logs remain to provide the necessary

Fig. 68 The largest native clubionids are the brownish ground spiders related to the Australian *Miturga*.

cover. A few species have extended their range from the typical forest to the sub-alpine regions where they are found sheltering beneath the loose stones but there is one species, *Uliodon hawkesi*, which has completely forsaken the forest for life among the loose rocks of the screes on the mountainsides in the Torlesse region of Canterbury. This species is strikingly different from its forest relatives because its colour is a very pale uniform brown, almost grey, and the legs are longer and very much more slender.

These spiders are certainly nocturnal hunters for, although the method they use to catch their prey has never been noted, they may be found prowling about on the forest floor and the trunks of trees at night. During the summer months the mother spider guards her huge, slightly compressed egg-sac in a sheltered spot beneath a stone or fallen log where she has constructed a roomy chamber, lightly lined with silk. In many parts of the North Island these spiders form the bulk of the prey of the large black and red spider wasps which are often seen struggling backwards as they drag their giant victims to the burrows they have prepared.

GNAPHOSIDAE

The bulk of the gnaphosids are dullish grey or black spiders which lead rather undistinguished and retiring lives. Remaining hidden within their silken retreats during the daylight hours, they only emerge at night to hunt for their prey. Many can move rapidly when necessary but their normal progression is quite slow although they have little difficulty in overcoming their intended victim whether it be an insect or another spider. If the victim is small, the attack will be direct but the

Fig. 69 The silvery *Anzacia gemmea* is the most widespread gnaphosid.

approach to a more dangerous adversary may be quite devious. The spider spreads her spinnerets out widely to produce a broad swathe of silk which is used to pin the victim down before she closes in for the actual kill. The name of this family has been subject to a number of changes and although Gnaphosidae is most commonly used today two other names, Drassidae and Drassodidae, may also be seen. Our commonest gnaphosid is *Anzacia gemmea*, a distinctive spider, about four and a half millimetres long with pale brown carapace and legs and a close covering of shiny hairs giving the abdomen the polished gleam of silver (Figure 69). The retreats are constructed beneath stones and unlike other gnaphosids may be found crowded together in large numbers. However like most of the gnaphosids the egg-sac, disc-shaped and covered with two sheets of fine silk joined along the margin, is guarded by the mother spider within the retreat. Many larger, grey or blackish gnaphosids live in a similar way under stones in open country, amongst the shingle of river beds or in the screes on the mountainsides but few of these have yet been given a scientific name. There is one group of species found in the South Island conspicuously different from these spiders, which although they have not yet been named, deserve some mention. These relatively large spiders, some six or seven millimetres long have two conspicuous black bands running down the whole length of the body which may be broken to form a row of patches on the abdomen. At first glance they may be thought to belong to *Clubiona* because they lack the clothing of shining hairs so characteristic of many of the gnaphosids but the anterior spinnerets are tubular and separated. The most striking of all the native gnaphosids is the attractive *Scotophaeus pretiosus* which as well as being one of the few forest dwellers in this family is also arboreal. *Scotophaeus* constructs its retreat beneath loose peeling bark, particularly favouring fuchsia and manuka so the easiest way to find it is to peel off bark from these trees.

The last two species we will mention are both introductions from Australia but these are the two gnaphosids which are most likely to be seen because as well as being large they are also found in settled areas. *Lampona cylindrata*, as its name suggests, is rather slender and cylindrical and can be up to fourteen millimetres long. The colour of the adults ranges from a dark grey to almost bluish black but there is always a conspicuous white patch at the tip of the abdomen. Two pairs of pale patches show on the abdomens of immature spiders so that they appear to be lightly striped. In Australia, this spider is said to be mainly terrestrial but in New Zealand it is usually found on the outside walls of sheds and houses where its slow deliberate movements are recognisable

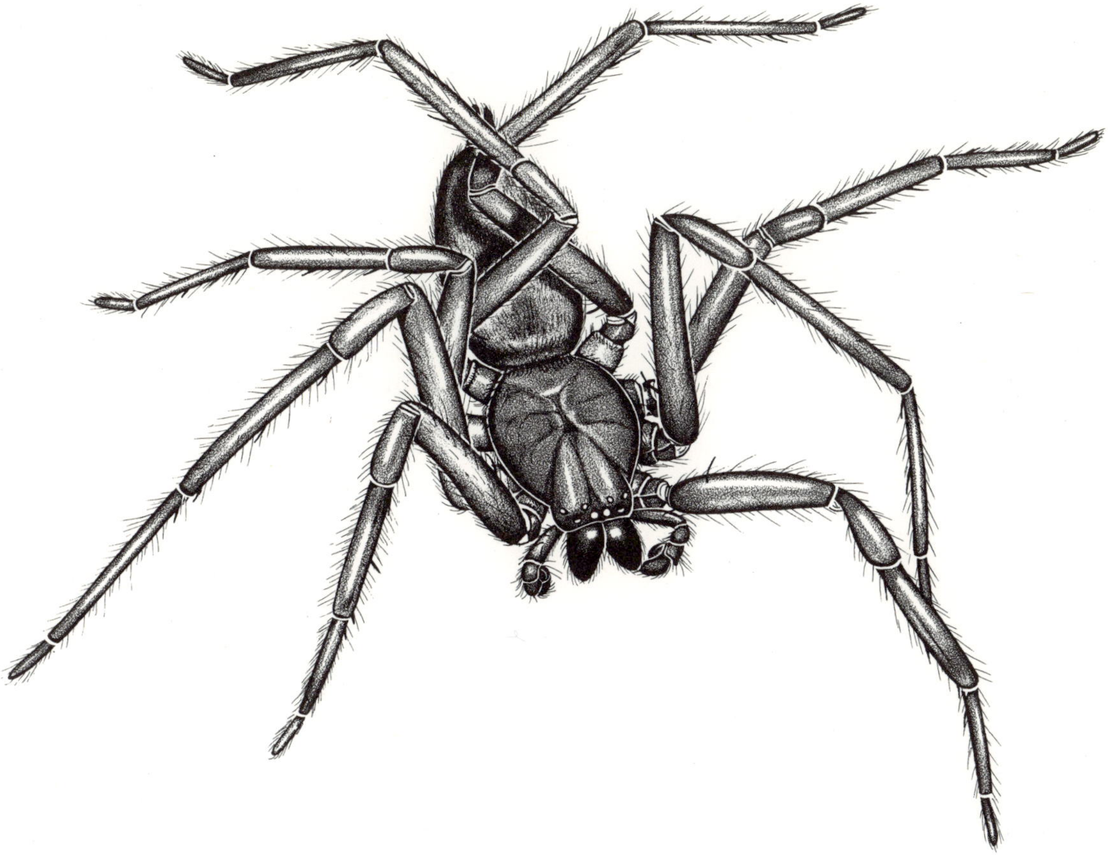

Fig. 70 The flattened bark spider *Hemicloea rogenhofferi* has been introduced from Australia.

from some distance away. There is still considerable doubt about its normal diet but the observations made by Russell Hutton in Napier suggest that the object of many of these forays on the walls of houses is the capture of other spiders and particularly it seems of its Australian compatriate the Black house spider *Ixeuticus martius*. Further observations need to be made in the North Island where this spider is common. The second Australian is a completely different spider which, with its flattened body and sideways directed legs, can easily be mistaken for one of the large crab spiders of the family Sparassidae. The greyish *Hemicloea rogenhofferi* seems to be governed by an urge to have both its lower and upper surfaces touching a surface as it scuttles from crevice to crevice (Figure 70). In its native haunts in Australia, its natural home is beneath loose bark. Although in New Zealand it is also found in this habitat and surprisingly enough shows a distinct preference for

Australian eucalyptus, it also inhabits fences constructed with overlapping palings and it is here that it is most often seen. It is interesting to note that, even with this preference for life beneath loose bark, the spider has not been able to find a home in our bush. This observation applies to all of the species introduced into this country but probably the very reason they have been able to reach New Zealand, in the main as stowaways, is that they can resist dessication, and associated with this is a preference for life in the drier areas. Unlike those of the gnaphosids, the eggs of *Hemicloea* are laid directly onto a thin sheet of silk spun on to the surface where the spider lives. They are covered with a smooth, papery sheet about three-quarters-of-an-inch wide so all that finally shows is a low dome where the mother remains until the spiderlings emerge.

CHAPTER EIGHT

Salticidae – Jumping Spiders

By far the most fascinating Arachnids are the Salticids or jumping spiders, so astonishingly anthropoid in posture and movements that one is tempted at times to ascribe human thought processes to them. From the moment the small, perfectly fashioned spiderlings emerge from their egg-sacs, the basic behaviour patterns are present, each succeeding moult adding only to their efficiency. The only major exception is associated with the development of the reproductive organs which reach maturity with the final moult. This perfects the mechanism which is triggered, in the presence of a male and female, into a series of courtship displays unequalled in the spider world.

Ranging in size from two to ten mm, these spiders in general are more conspicuous in the North Island. Their frequently bright and attractive colouration is broken up by bands, stripes and speckled patterns while their bodies and legs are adorned with special hairs and spines. Such eye-catching features are significant in courtship and, combined with other characteristics, make recognition of this group easy. The distinctive rectangular cephalothorax is capable of swivelling at right angles to the abdomen, thus providing the action necessary for the visual efficiency of the eight eyes. Four of these are large and facing forwards, prominently situated on the squared front of the carapace, while a tiny pair of laterally directed eyes forms a row behind them. A third row consists of another pair of moderate-sized eyes whose range extends both upwards and backwards (Figure 71). It is believed that the four eyes of the second and third rows are specifically concerned with the detection of movement and all our observations confirm this. On the other hand, the large middle pair of front eyes is said to be capable of receiving an image enabling it to distinguish prey, a prospective mate or an enemy. To some extent this may be so but the sense organs of spiders have not all been studied in detail and it must be remembered that the lives of invertebrates are

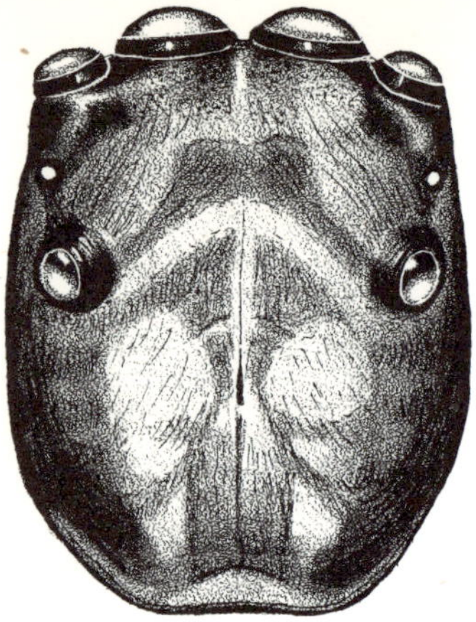

Fig. 71 The carapace of a salticid.

largely governed by a set of complex instinctive reactions to specific stimuli beyond the normal range of human detection. Nevertheless, it is evident that these eyes do possess powers beyond those of the others and that they are concerned with the recognition of shape and size together with distance estimation and the perception of light and dark areas. This ability has enabled the salticids to abandon their snares and simultaneously evolve the specialized movements associated with hunting and mating.

It is no exaggeration to say that these spiders jump, hop, skip, crouch, stretch, run, leap and dance; no choreographer could wish for a more varied repertoire. Such antics vary between species and have led to a broad separation between hoppers, jumpers and runners, all of which are able to jump but some do so only under specific circumstances. The front pair of legs, heavily built and in some species noticeable larger are not, as may be thought, used for grasping prey although they are sometimes used for holding the victim still. Instead, evidence suggests that their role is primarily that of a signalling device in addition to possessing chemo-tactile sensitivity. Spiders with one or even two of their front

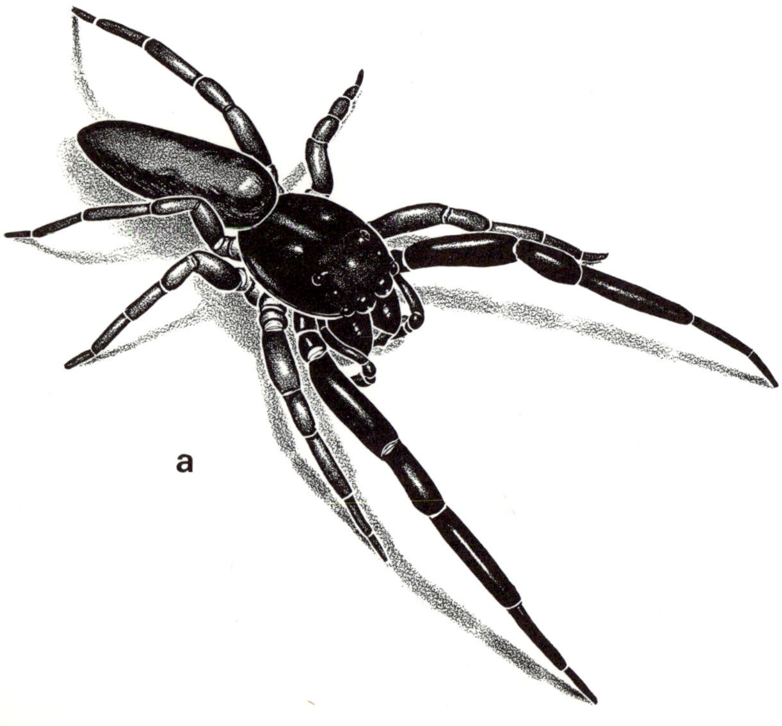

Fig. 72 *Trite planiceps* beginning to stalk an insect. a) The prey is detected at the side and the spider stops. b) The cephalothorax is swivelled so that the front eyes are peering at the prospective victim. c) With the body straightened the spider is ready for the final advance.

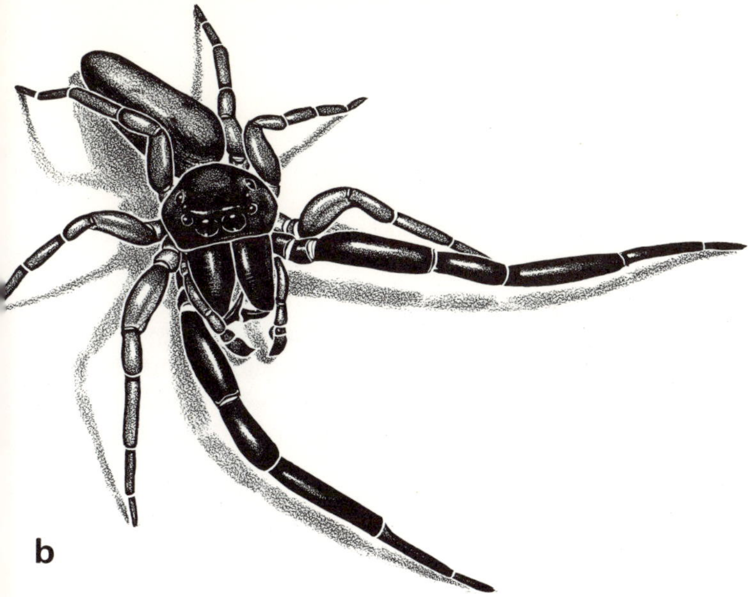

b

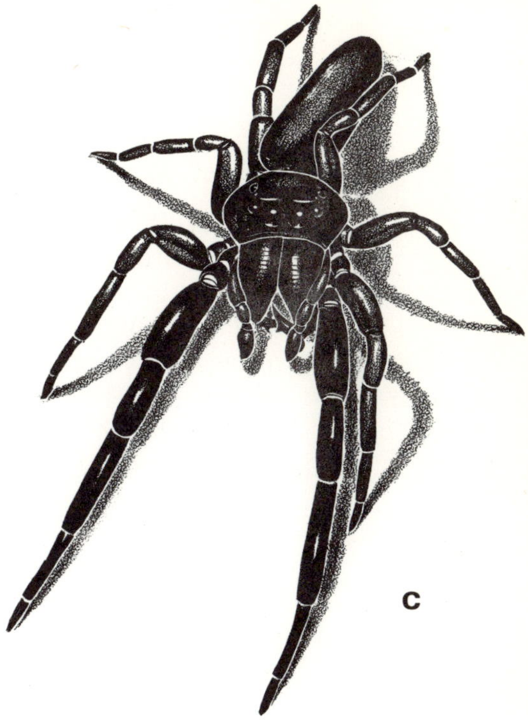

c

legs missing are certainly not inconvenienced in either their normal walking or jumping activities and prey-catching proceeds with as much vigour as usual. It seems extraordinary, therefore, that with the obvious modification of the front legs they are not in fact used for jumping, but rather that this action is performed by the relatively simple and apparently unspecialized third and fourth pairs of legs.

Whatever the mode of progression it is marked by a dragline of silk laid down by the spinnerets. Not infrequently, this jumbled network may be gathered up and actually eaten by the spider. Apart from this, the only other use of silk is for the construction of retreats where the spider rests at night and also moults, and to cover the eggs after they are laid. This release from the dependence on snares for prey-catching enables the spider to go further afield in its search for food. In addition, one factor seemingly common to this group is its preference for a comparatively dry and warm environment. Unlike lycosids and others, these spiders should not be kept in containers with moist cotton wool but will remain perfectly content in a dry glass jar away from bright sunlight. In these containers they will even live for long periods without food, a fact which is remarkable when you consider that they are also without moisture.

There is much to be done on the study of these spiders in New Zealand and even overseas information is sketchy. The most important work on this group has been that of Jocelyn Crane whose detailed account of the comparative behaviour among several sub-families of salticid spiders should surely inspire someone in New Zealand to do likewise.

Our most widely distributed species is the *Trite auricoma* – a squat, golden-brown spider found in most parts of New Zealand. This is broadly grouped as a runner because it is comparatively slow-moving, and jumping is reserved largely for prey-catching, although it may occasionally jump across a barrier. Blending perfectly with a background of soil, leaf-mould and pebbles, it can remain virtually unseen until it moves. The male is distinguished from the female by a yellow clypeal band between the frontal eyes and the chelicerae.

Employing a specialized combination of movements, this spider makes full use of its fine eyesight to stalk and catch its prey. If a fly alights within about twelve inches in front, the spider is immediately aware of it. First, there is a slight shifting and stiffening of the legs, then the hind legs are drawn in and a slow, stealthy stalking takes place, the spider sliding imperceptibly towards its quarry. Even if no further movement of the fly takes place it does not matter; his death is assured as the last two inches are covered with a lightning leap, and outstretched fangs pierce the victim with a tenacious grip from which there is no escape. If the fly alights behind the spider, the movement is detected by the posterior pair of eyes. Instantly the spider swings round to face the prey and then goes through his stealthy stalking motions again in the same way as illustrated for *Trite planiceps* (Figure 72). Sometimes the fly delays the moment of destruction by moving. It avails him nothing, for now the spider postpones his stalking and follows his victim around with watchful gaze, turning his cephalothorax with a swivelling action until it is at right angles to the abdomen. Quietly, imperceptibly, the abdomen slides into line with the cephalothorax, the stalking begins again and ends in that final, inescapable jump.

There is no doubt that this spider is able to determine the difference between enemy, prey and a possible mate with uncanny accuracy. Vision is unquestionably involved. Yet a male will use the same initial signal in the presence of either male or female and only when he is facing the female at a distance of not more than two inches will the pattern of his leg-waving change. In addition there are variations between threat and inter-male display. The female on the other hand catches her prey in an identical fashion to the male, remains passive if she is sexually receptive to a mate but if she is not, or if confronted by an enemy, her threat posture is instantaneous and forbidding. The front two pairs of legs become stiffened and straight, the carapace is raised, the abdomen is bent at right angles to the thorax while firmly touching the ground, the palps are rounded and chelicerae extended although the fangs are kept recessed. Perhaps it is only our imagination but even the eyes seem to have a baleful glare. Comical though it appears to us, there is no doubt of its effectiveness for, thus confronted and outstared, the enemy turns tail and flees. A male treated thus forgets his intentions and leaves the scene in great haste. It is common belief also that the female devours the male upon the least pretext and that if he is not quick to depart after mating he is promptly eaten. We have never seen this even when we have had the two spiders in a container from which there is no escape, nor have we ever seen the slightest attempt at cannibalism.

Two females brought face to face will at once adopt the threat stance which is always accompanied by side-stepping, first right and then left although this seldom brings the two antagonists any closer. This movement serves to draw the attention of one to the other effectively, saying, "Look, I am threatening you". Where two females are concerned, the stance and movements are continued until the "weaker" one gives way, although what determines this weakness it is difficult to see. The confrontation of two males of *auricoma* is an interesting one and so far has not been observed in any other species in New Zealand (Figure 73). The spiders swing round to face each other and first one and then the other will raise the front legs sideways at an angle of about forty-five degrees. These legs are then waved up and down, each rhythmical movement being accompanied by side-stepping. At the same time, the spiders approach cautiously with their eyes firmly fixed on each other. In addition each cephalothorax is raised high with the front two pairs of legs tiptoeing, abdomen bent firmly to the ground and palps rounded. Finally the tips of their front legs touch while in the raised position, then drawing even closer together both fangs and palps are brought in contact. This position is held for several seconds after which the two break apart and take no further interest in each other for about five minutes. Under normal circumstances, this would give ample time for them to have left the scene and it is unlikely that their paths would cross again; in a confined area however this procedure is repeated fairly regularly but gradually the entire sequence is cut short. The evolutionary value of such a ceremonial tournament would be to provide for the release of aggressive tendencies without any loss to the species.

It is amusing, to say the least, to watch the male's reaction to his own image in a mirror (Figure 74). The normal inter-male behaviour is repeated until he actually touches the glass surface. Frantically he strains at the mirror in effort to make contact, even climbing up on to it or reaching with one leg behind. Slyly he turns away for a few steps, then swings round abruptly to see if his opponent is still there. He

Fig. 73 Two males of *Trite auricoma* stage a mock battle. a) The cephalothorax is raised and the front legs are curved in front. b) As neither retreats the front legs are extended to the side, the fangs are extended and they advance towards each other, with the body raised high. c) The front legs touch and they advance closer but as soon as the chelicerae touch they will break away.

Fig. 74 Aggressive behaviour may even take place when a male sees his reflection in a mirror. a) The initial bent front legs stance. b) The advance with extended front legs.

stands stock still, with astonishment seemingly written all over his face. After several futile attempts to touch his opposite number, the spider finally gives up and actually shows a strong tendency to ignore the mirror at later trials.

It seems rather a shame to spoil such a pretty little story and anyone who takes the time to repeat that experiment with one of these spiders will find it hard not to let his imagination run away with him too. It seems evident that the initial stimulus between two males, or a male and his mirror image is a visual one but the final consummatory act is probably a chemotactile one. This explains the efforts which are made to have contact in order to complete the chain of co-ordinated behaviour patterns. When he turns away as if trying to take the other by surprise, it is the posterior pair of eyes which detects his own movement in the mirror thus stimulating a reaction, that of swinging round to face the intruder, which we have already seen in other situations. It is so temptingly easy to place our own human interpretations on what we see that it can be hard to realize the role instinct plays in many instances. On the other hand it does seem that the spider has 'learned' that there is 'something fishy' about a mirror image when he refuses to be tricked at a later stage. Scientifically, this is usually referred to as habituation, or in even more technical terms, that it is the result of action-specific exhaustibility!

Courtship displays have been the subject of much interest and analysis, particularly amongst birds whose attractive dances were once thought to be of value in the selection of a suitable mate. Indeed the Peckhams who first described such performances by the Salticids, also held this view. Now it is considered that they have evolved as a means of species recognition, thus facilitating mating and preventing crossbreeding. Such ritualized behaviour is in fact highly differential and it is significant that the appropriate displays bring immediate response from members of the same species (Figure 75). So reliable is this that we have frequently employed it as a method to determine whether two spiders possessing some superficial variation really do belong to the same species.

Females are by nature stay-at-homes so we were fortunate to capture one recently, thus enabling us to observe and photograph both courtship and mating behaviour. Males we had in plenty for they appear to wander in search of a mate; almost every day we would catch one on the walls of our kitchen. The initiative is taken immediately by the male while the female, if she is receptive, remains passive in a crouching position. Although the female may not accept the male at once, her threat stance if genuine is enough to deter the stoutest male heart. As in the first stage of inter-male behaviour, the front pair of legs is raised sideways at an angle of forty-five degrees and waved up and down while the spider side-steps, two to the left, two to the right. Even if the spiders are not facing each other to begin with, this action brings them smartly round

53. *Diaea* sp. Male. Family Thomisidae. Dunedin. Length 3 mm.

54. *Diaea* sp. mating. Family Thomisidae. Shingle Creek. Length female 2.5 mm.

55. *Diaea* sp. Female guarding egg-sac within retreat. Family Thomisidae. Christchurch. Length 2.5 mm.

56. Arboreal Jumping spider. Male. Family Salticidae. Erua, Length 4 mm.

57. Tree-trunk jumping spider. Female. Family Salticidae. Catlins. Length 5 mm.

58. Tree-trunk jumping spider. Female. Family Salticidae. Dunedin. Length 5 mm.

59. Riverbed jumping spider. Female. Family Salticidae. Poerua River. Length 4 mm.

60. House jumping spider. Male. Family Salticidae. Masterton. Length 4 mm.

61. House jumping spider. Female guarding egg-sac. Family Salticidae. Feilding. Length 5 mm.

Fig. 75 During courtship a male salticid uses his front pair of legs to signal to his mate.

face to face. The male continues his movements somewhat jerkily as if to emphasize them, all the while keeping his eyes fixed firmly and judiciously on the female. When he is within two or three inches of her, his leg movements change, providing of course that the object of his intentions is still there.

Sometimes she moves away so that the male is obliged to follow her, gesticulating wildly. However, once more frontally positioned, the front legs are raised forwards still waving up and down at about the same angle. They touch the ground in front of the female several times while the male now moves cautiously towards her. She then raises her legs and touches his. Still using his legs, the male moves over on top of her at which stage all show of resistance ceases and she remains quiescent until mating is finished, a process which we timed at forty minutes. The male changes his position from the left side of the female, using his left palp, to her right side using his right palp, after about twenty minutes. During one experiment we interrupted the male at what we felt was about half-time and then when in about five minutes he located the female again he moved very rapidly through a couple of forward leg-raising actions and then directly to the female to use his left palp. Unfortunately, we were not able to use an entirely new female to discover what would have happened in that instance, although it seems reasonable to conclude that the use of the left palp was simply the result of the lack of any seminal fluid in the right one. Our observations have been restricted because since then we have found only isolated specimens.

After mating, the females go into seclusion somewhere, and this is a most successful operation too, because during the winter months we have not located any around our property – neither males nor females. Yet in the spring the females are discovered under stones and planks guarding their egg-sacs within a retreat. At this time there is no sign of mature males and it is not until January that they appear on the scene again. So, on the basis of our house and garden investigations, we can only conclude that the female mates in Autumn, virtually hibernates over the winter and lays her eggs in the spring. From then on young spiders in many stages of immaturity are found; but we find it hard to explain the appearance of mature males in January unless they too have survived the winter.

More spectacular in both appearance and activity but less widely distributed is another member of this genus, *Trite planiceps*. Moving with great rapidity along the under surface of the leaves of flax and cabbage trees, this spider can bridge a gap of up to two feet with both grace and precision. One of the largest of jumping spiders in New Zealand, it is distinguished by a shining ebony carapace and large black forelegs together with a slender abdomen longitudinally marked by a central yellow stripe. The remaining three pairs of legs are transparent amber in colour. Possessing both the highly characteristic alert and swivelling positions of other Salticids, this species demonstrates by its agility the peak of arachnid visual acuity. Prior to jumping, both front legs are extended forwards in the direction of the anticipated leap, an action seemingly associated with some form of distance estimation – a bit like sighting along the barrel of a gun (Figure 77). Almost as disconcerting, too, we found, for as soon as we came within a couple of feet, the subject of our observations would land on the camera, pencil, magnifying glass or even our noses. Fortunately perhaps, this spider is capable of an even more extraordinary feat, that of returning almost instantaneously to the spot he has just left, should his objective prove unsuitable or undesirable.

In order to test his ability to estimate distance, we placed the spider on a piece of glass held at varying heights above the ground, whence he would jump quite readily until a height of about two feet was reached. At this point he would poise himself on the edge of the glass, partly raise his legs in preparation for the jump and then hesitate, move along

Fig. 76 *Trite planiceps* guards her row of egg-sacs within a curled up flax leaf.

to another spot and try again. If the glass was lowered slightly he would make his leap at once. Constantly on the move, he would vanish if we took our eyes off him for a second. Yet recognition of the size and nature of prey must be largely instinctive, for we have never found one making the mistake of lunging for a blowfly or beetle which would be too big to handle.

At rest this spider lies in a lengthwise position with the first two pairs of legs stretched out in front and the third and fourth extended behind, a fact which should have led us to his hiding place sooner than it did. Our sweeping and beating disclosed not a trace but careful unrolling of the cabbage tree leaves and flax, soon located them in all stages of growth (Figure 76). There, too, were the silken retreats, the moulting hideaways and the discarded skins. There seems no reason for this spider not to be more widely distributed than it is but we have never found it further south than Christchurch.

There is no particular distinction between the male and female as in *auricoma*, except that the female is a little shorter and a little plumper than the male. Their display activities vary considerably, with the female assuming the role of a leg-waving male under certain circumstances.

Fig. 77 *Trite planiceps* catching a fly. a) When near enough to bridge the gap the front legs are extended out in front. b) The fly is immediately transfixed with the fangs.

During the courtship, however, she remains passive while the male performs a sideways leg-waving action, sidestepping as he approaches. He displays only briefly to his mirror image and yet we have also obtained a partial reaction to a cardboard model which pictured the leg position of a spider in front view. In two instances we found a fully mature male closeted in a retreat with a nearly mature female which moulted within two days. With the male thus remaining in the vicinity of an about-to-moult female, successful mating is virtually assured. At least four egg-sacs are laid within a silken retreat arranged in a row inside the rolled up flax and there the mother stays until the spiderlings emerge.

An excellent example of the kind jumping of spiders which we call "hoppers" is a species which is found in large numbers on the outside walls of houses in the North Island. Even though some sixty names have already been bestowed upon New Zealand species of salticids we have not been able to find a description which would cover this species. We suspect that it was introduced from Australia although,

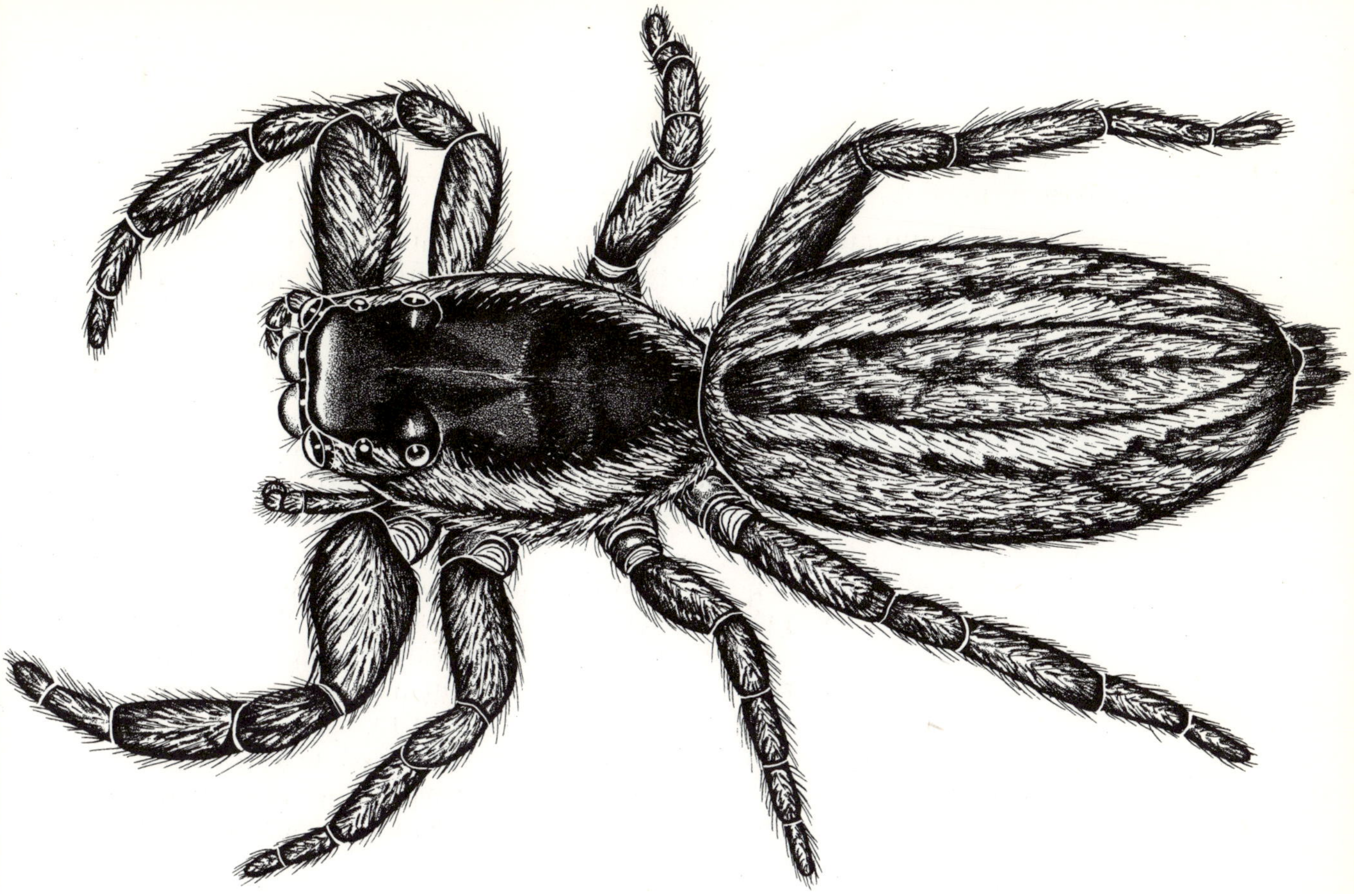

Fig. 78 The flattened *Holoplatys* is at home in narrow places.

as we will mention later, other related species which are undoubtedly natives are found in various habitats throughout New Zealand. For the present, we will refer to this species as the House Hopper. This delightful little spider may be found sunning itself on almost every house in the districts where it occurs; our own observations were made in Feilding and Foxton where it is extremely abundant. These hairy, brown jumpers are patterned attractively with dark and light areas, giving them a speckled or striped appearance. They build their silken retreats in the cracks and crevices on the walls and when the sun has gone, even temporarily behind a cloud, they vanish within. When kept in captivity they immediately begin to spin their retreat. We found that the best way to bring them out was to place the cage out in the sun. Many an hour was spent just watching them catch their food on the vertical walls which they are able to traverse with the aid of the special bunches of hairs beneath their feet but always with the added insurance of a thin but strong dragline which they attach to the surface as they move. To us, the most extraordinary feat seemed to be their ability, after stalking a fly to within about three inches, to defy gravity with a horizontal leap and land plumb on the doomed fly blithely grooming itself on the wall. It seems that the sheer momentum of the powerful jump projects the spider on to its objective before the pull of gravity takes effect; otherwise we would have to postulate that the spider has learned to adjust the direction of its aim according to its height from the ground and this might be crediting it with a degree of learning ability it does not have. There is little doubt that this spider is among the most keen sighted spiders in this country, although we have at times found that it will leap upon the head of a nail if this projects out from a surface, thinking it to be prey.

Perhaps the most exciting period during our leisurely peering on walls was the brief observation of a pair doing their courtship dance, still, mind you, on the vertical wall. All of the score or more species we have watched at one time or other have used the front pair of legs to carry out their elaborate series of signals but the male of this species lifts up his third pair of legs so that it is well above his body, while at the same time drumming his palps on the surface and weaving from side to side (Figure 79). Some time we hope to continue our holiday observations of this fascinating spider.

Numerous small 'hoppers' are found on and beneath the bark of forest trees but those most closely related to the House Hopper live in more exposed places. Along most pebbly beaches in New Zealand, well above the high tide mark are found a number of species of these large, black or dark grey salticids of which only one, known as *Marpissa marina*, from along the Otago coasts, has been named.

The remarks of Goyen who described this species in 1891 are of interest.

> This little spider is found along the coast of Otago, on cliffs and rocks just above, at, or just below the high-water mark. On these rocks and cliffs are found in great abundance three species of Diptera, which the spider resembles in colour and mode of progression. So striking is this resemblance that for some time I mistook the spider for the fly. In moving along the face of the rocks it runs very briskly for an inch or two, then stops, rubs its palpi just as a fly rubs its fore-legs, examines the surface of the rock within its view, then again runs briskly for an inch or two, again stops, rubs its palpi and "spies out the land"; and so on until it comes within view of its prey – or what it takes for such, for it seems to have some difficulty in distinguishing members of its species from the flies on which it lives. It now advances exactly like a cat approaching a bird, stealing forward at a pace so slow that one can scarcely see it advancing, until within about an inch of its prey, when it springs swiftly upon the unsuspecting fly and

Fig. 79 The male House Jumper is one of the few New Zealand salticids to use its third pair of legs during courtship.

> pinions it with its falces and palpi. Within springing distance its aim is so sure and its bound so swift that the fly has no chance of escape. Having captured its prey it generally carries it to a depression in the rock, and there, if undisturbed, sucks its juices at leisure. Half the spiders one sees are in hiding with a fly in their jaws. Very often the fly is much larger than the spider; but the latter if discovered by another spider, makes off with the captured fly with the greatest of ease. It will even leap across a crevice an inch or so wide, and if its covetous brother is persistent in his pursuit, to escape him it will spring into space and hang, head downwards, suspended by a strand of the web, but never parting with its prey. In every case I observed the pursued was successful in eluding the pursuing spider."

The general observations made by Goyen can be applied to a greater or lesser extent to a whole series of related dark coloured and seemingly drab species which inhabit the shingly river beds and the bare rock faces of the mountainsides high above the upper edge of the forest. All of these spiders seem to show greatest activity while the sun is shining, for this is when they may be seen hunting.

This account of the jumping spiders would not be complete without mentioning the oddest members of the family, the curiously flattened, grey species of *Holoplatys* which look as if they have been pressed between the pages of a book as one might press flowers and leaves (Figure 78). This body form provides the clue for the habitat they prefer, the small cracks and crevices in and under the bark of trees, but man-made constructions such as overlapped paling fences also provide an effective home and it is here many of these peculiar spiders are seen. Closely blending with the weather-beaten wood on which they lurk, they exhibit a special behaviour pattern known as thigmotaxis, a compulsion to have both dorsal and ventral surfaces of the body in contact with a surface, which combined with a curious sideways gait enables the spider to retreat immediately into any cracks and crevices available.

In spite of the large bulky forelegs and ungainly appearance, these spiders can catch a fly as easily as others of this family. They show the same behavioural reactions as *auricoma*, raising the body to observe prey, stalking and jumping but the front legs play no part in this activity. Although squat, the cephalothorax still turns at right angles to the abdomen, thus indicating normal visual ability.

CHAPTER NINE

Crab Spiders

Crab spiders are always somewhat flattened but this effect is enhanced by the position of the legs which are directed sideways to the body and usually kept close to the ground. All are vagrants and are considered hunters although they have only poor eyesight. Prey is usually caught as they wait motionless until the strong front pair of legs can be brought into action, clasping and then crushing the victim to their fangs. The families Sparassidae and Thomisidae are found in New Zealand, although only the true crab spiders of the family Thomisidae are really native in this country.

THE SPARASSIDAE

Most members of this family are huge and the name which has become accepted for them by common usage is Giant Crab spider. None of these spiders is native but the Banana Spider *Heteropoda venatoria*, as might be expected from its name, is continually brought in with bananas and even after fumigation it often manages to survive (Figure 80). In Dunedin alone, on an average, five of these spiders are brought in for identification each year and in spite of their size have long since lost their newsworthy status of a photograph in the local newspaper. Many are females carrying their grey, lens-shaped egg-sac beneath the cephalothorax. As each of these egg-sacs contains hundreds of eggs, it may be wondered why the banana spider is not by now a familiar sight but fortunately for the peace of mind of the housewife, the young do not thrive in our climate, and the spiderlings rarely live beyond their second moult. In most tropical countries these spiders are a familiar, and often welcome, sight as they scuttle over the walls, capturing and eating many obnoxious insects including cockroaches which form their favourite diet.

Fig. 80 The tropical banana spider *(Heteropoda venatoria)* is often brought in with tropical fruits.

There is one member of this family which is now firmly established in the Auckland district. The huge and hairy Australian Triantelope *Isopeda insignis* made its first appearance in the Avondale suburb of Auckland some twenty years ago and is now called the 'Avondale' Spider (Figure 81). These harmless spiders with their two-inch leg-spread are a familiar sight in many Australian houses where they are usually tolerated and sometimes appreciated for their diligent hunting of household pests, and it may well be that some day they may enjoy a similar role in this country.

THE THOMISIDAE

True crab spiders of the family Thomisidae are plentiful on the foliage and trunks of trees and shrubs, spending most of their time above the ground but even so they are not seen as often as other spiders. Blending closely against the background of the habitat they prefer, these spiders represent one of the peak adaptations in camouflage. At present, three genera contain the six species which are recognised but this is one of the groups to which scant attention has been paid, and in fact some scores of species are to be found which eventually will be placed in four or five genera. However, on general appearances, all our species both described and officially unknown can be divided easily into the three genera we use at present and each of these groups occupies quite different ecological niches.

Only a single species of *Cymbachina* (*C. albobrunnea*) is known and this from no more than a dozen specimens. It is probably fairly common, but no one has looked carefully for it in the places it prefers. This conspicuous

Fig. 81 The Avondale spider *(Isopeda insignis)* Male. This common Triantelope from Australia is now well established in Auckland.

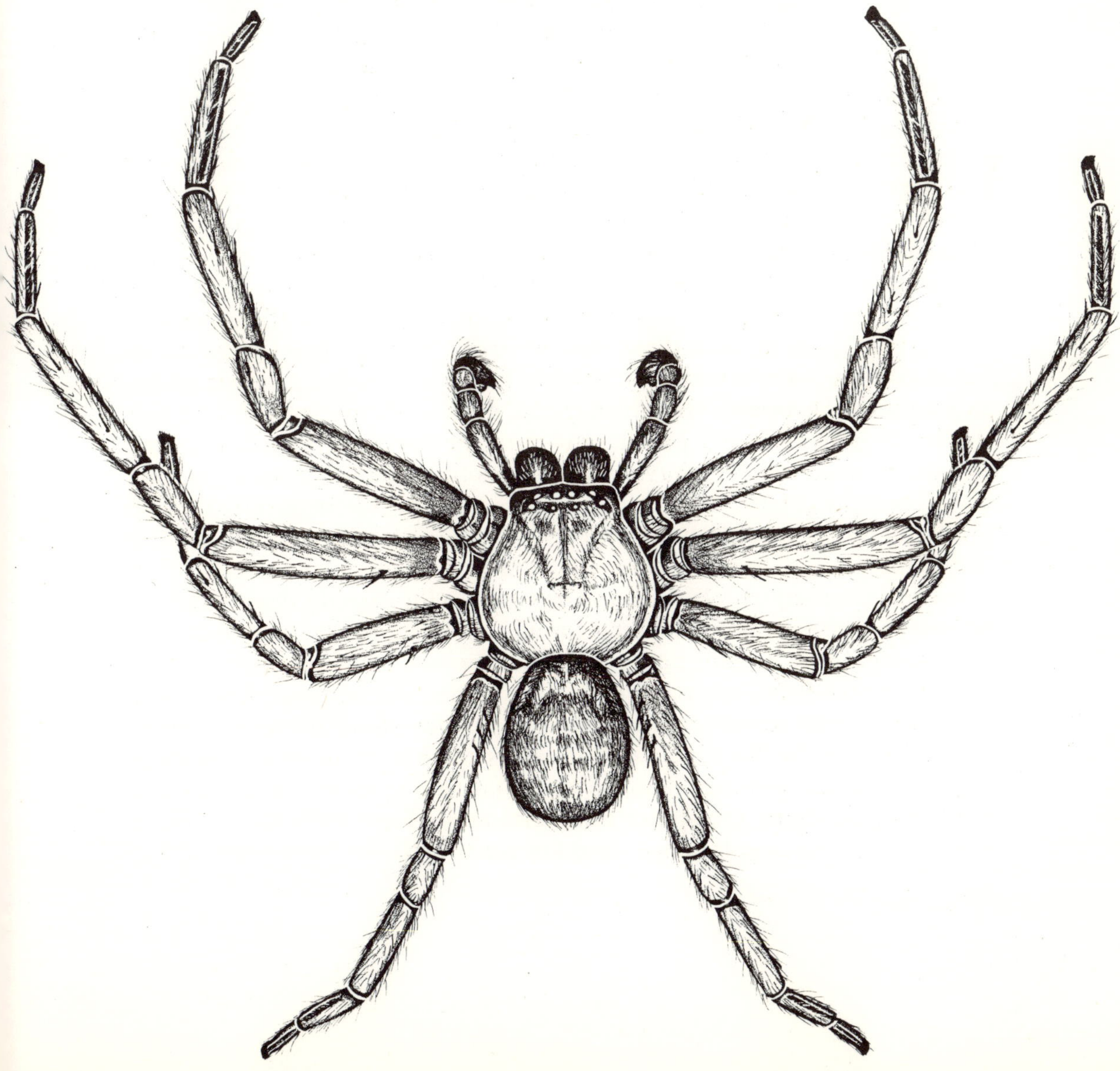

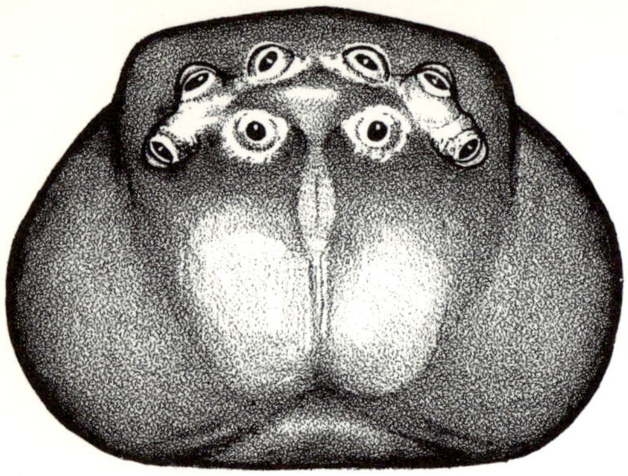

Fig. 82 *Diaea* carapace (Thomisidae).

little spider with a mottled black and white body and legs makes its home along the black and white lichens on the trunks of trees where it is impossible to detect unless it moves.

The grey to reddish-brown crab spiders with the extraordinary blunt-ended abdomen are all placed in the genus *Sidymella.* Among the twenty or so species, the carapace and abdomen are sometimes smooth but more often granular and leather-like. A cursory glance indicates the possible habitat and indeed these spiders are found where they match the background perfectly on the bark of trees or occasionally on the forest floor. There they wait patiently for their quarry to wander close enough to be scooped up with the spiny pairs of legs. Apart from the dull and drab colouration which is readily accounted for when one finds them so admirably concealed, the peculiar cut-off shape of the end of the abdomen has raised many speculations because only rarely do unusual modifications persist in animals if they do not have some use. While we may not have been able to provide a complete answer to this question, we certainly found that these spiders make use of their shape in an extraordinary way. When most spiders become stranded on their backs, as frequently happens, they manage to push with the legs on one side until they gain sufficient purchase to turn themselves over sideways and so right themselves, but *Sidymella* achieves this result in its own peculiar way. Its main defence when disturbed as in most Thomisids, is to drop on the end of a thread of silk where it dangles with legs outspread, but *Sidymella* always seems to land on its back. As soon as it reaches a firm substrate it severs the thread and remains lying upside down with its legs curled over its body, looking for all the world like a particle of dirt. The spider may remain completely motionless for some minutes but eventually the front two pairs of legs are stretched out and then drawn back under the cephalothorax. After a further short pause, the front two pairs and the hind pair of legs are stretched to lever the front of the body off the ground. Because the front two pairs are very long and the hind pair short, the spider now rests at an angle with the abdomen supported on its angular edge. A sharp push with the front pairs of legs raises the body to a vertical position steadied by the fourth pair of legs but with the remaining three pairs free, outstretched to balance it. This acrobatic stance with the spider resting squarely on the hind portion of the abdomen seems perfectly stable; here it may remain for some time, until, by a flick of the legs to the front of the body, the centre of gravity is changed and it topples over to land the right way up when it is ready to move away (Figure 83, a–i). The first time we saw this complicated manoeuvre we thought it must surely have been an accident but no, after repeated experiments, we now know that this happens every time. With this unusual preview of the habits of these spiders we looked forward to recording some spectacular aspects of their mating behaviour but this we found most prosaic.

Although the male is much smaller than the female, his approach seems devoid of any preliminary behaviour to ensure that his advances will be accepted. He moves straight up to the female as she rests motionless and lightly tapping her with his front pair of legs creeps immediately on to her back. There, while drumming with his palps, he moves around until he faces the end of her abdomen and after a short pause creeps under her abdomen so that he is facing the same way as his mate, but upside down. After applying his palpal organ to one side, he moves back on to the top of her abdomen and creeps beneath the other side to apply the other palp. He may cross from one side to the other four or five times and up to half an hour may elapse from start to finish. The end comes as abruptly as it starts as he just climbs off and departs.

All of the brightly coloured crab spiders which live on the foliage are at present placed in the genus *Diaea*. While these spiders are predominantly green, the group to which *Diaea albolimbata* belongs is also decorated with black, and occasionally red, patches and bands. The front two pairs of legs are extremely long (Figure 85). The second major group which is characterised by *Diaea ambara* is a uniform green or sometimes yellow while the front legs are relatively shorter. After reading the delightful account by W. S. Bristowe in his book *The World of Spiders* about the mating behaviour of similar looking species in England we hoped that we might find the same habits among some of our species of Diaea, and in fact we did.

We decided to study a bright green species related to *Diaea ambara* which we found living in large numbers on the low shrubs in Central Otago. It was not until dusk when we noticed a male whose seemingly purposeful wanderings suggested to us that he was seeking a mate. Eventually he reached a leaf on which a female was resting with her front pairs of legs outstretched in the typical hunting pose. She made

Fig. 83 *Sidymella* takes advantage of the blunt end of its abdomen to turn itself over. a–c The front two pairs of legs are drawn back under the body. d–e The two front pairs of legs are extended lifting the body until with a sudden jerk the spider can rest upright on the base of the abdomen. g–i). With a forward flick *Sidymella* topples over to land the right way up.

a

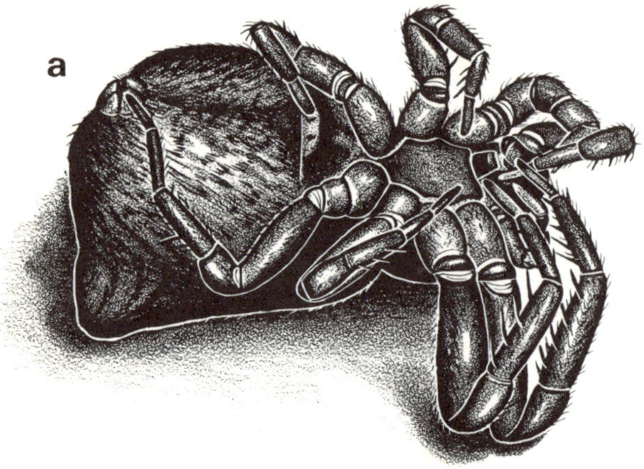

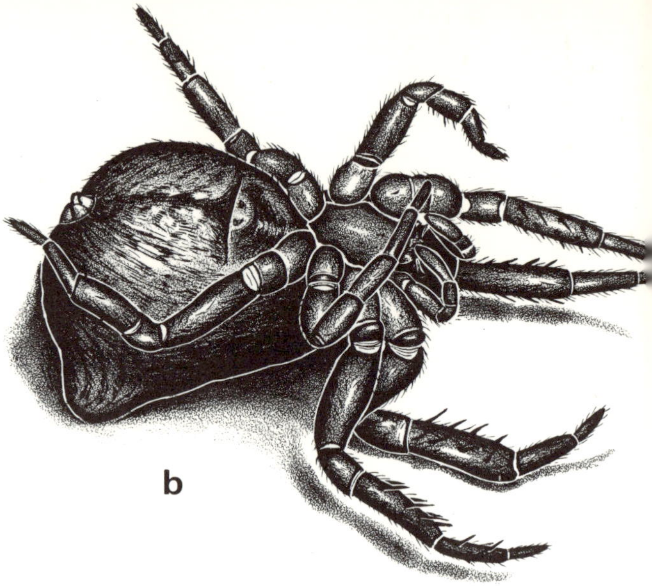

b

c

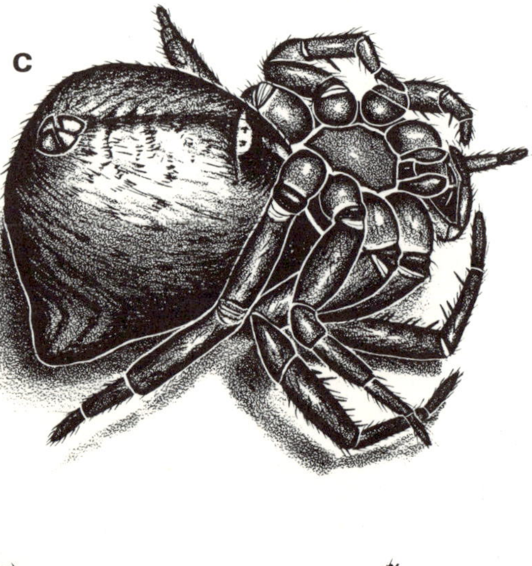

d

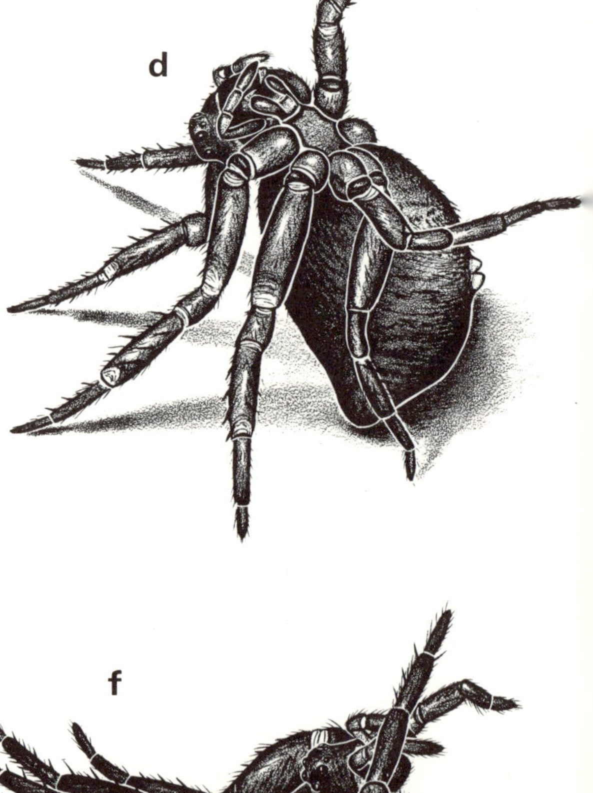

f

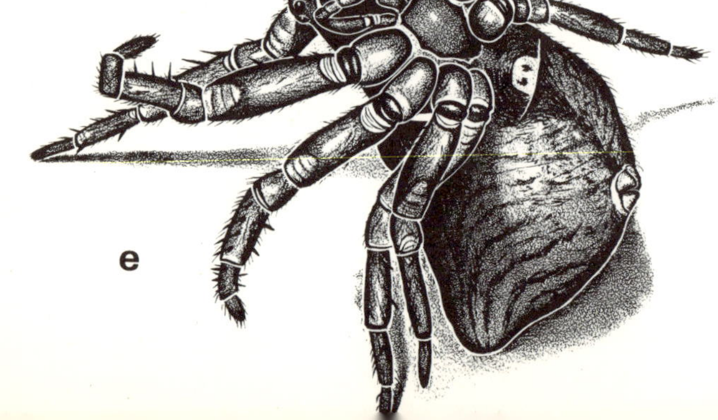

e

g

h

i

Fig. 84 The small male of *Sidymella* creeps over the female and reaches beneath her abdomen to insert his palp.

Fig. 85 *Diaea albolimbata* is found on plants.

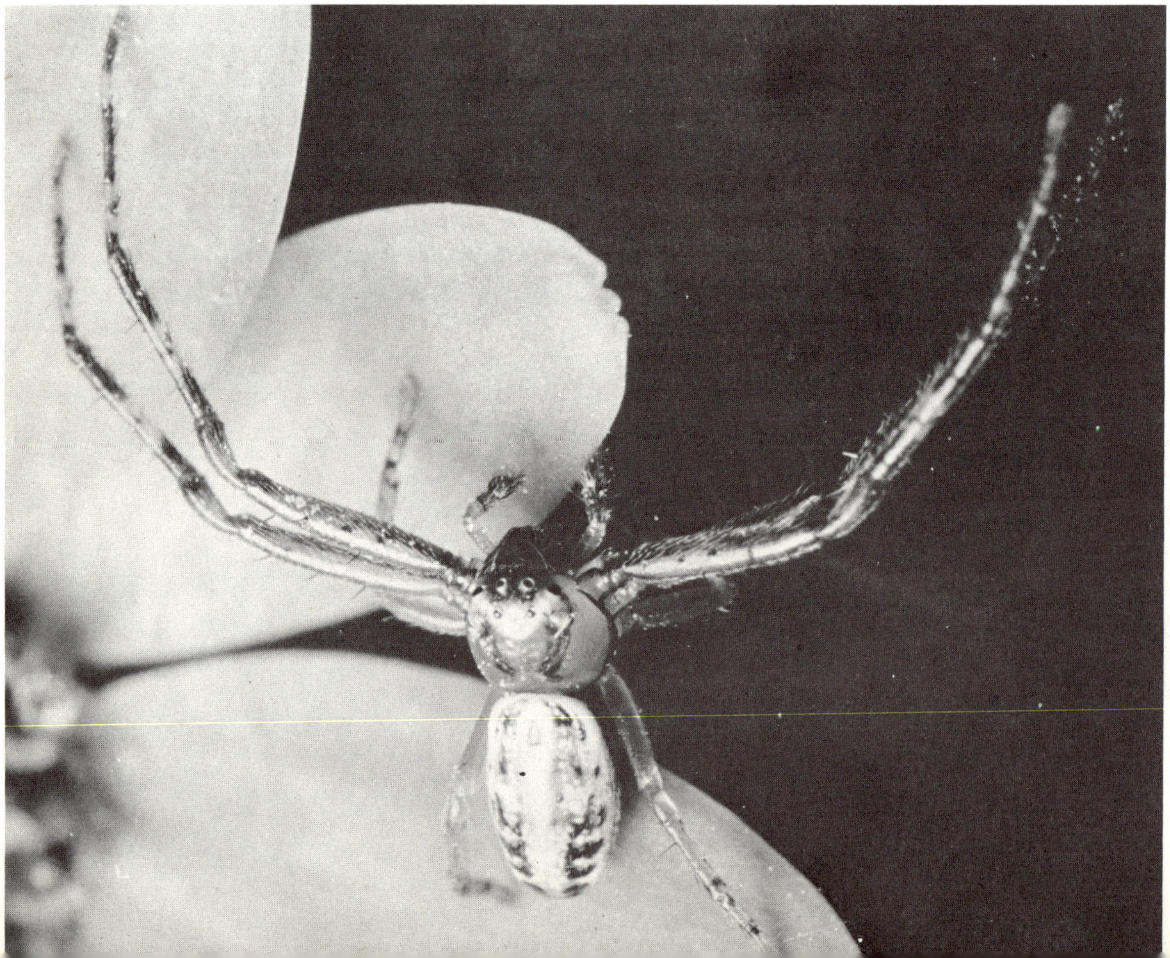

62. Top left. *Holoplatys* sp. Female. Family Salticidae. Takitimu Mountains. Length 5 mm.
63. Top right. *Trite bimaculosa*. Male. Family Salticidae. Waipori Gorge. Length 5 mm.
64. Left. *Trite planiceps*. Mating. Family Salticidae. Christchurch. Length 10 mm.
65. Below. *Trite auricoma*. Mating. Family Salticidae. Dunedin. Length 9 mm.

66. *Dysdera crocata* in retreat. Family Dysderidae. Dunedin.

67. *Dysdera crocata*. Female. Family Dysderidae. Foxton. Length 14 mm.

68. *Segestria* retreat. Family Segestriidae. Feilding.

69. *Ariadna* retreat. Family Segestriidae. Nelson.

70. *Ariadna* sp. Female. Family Segestriidae. Portobello. Length 11 mm.

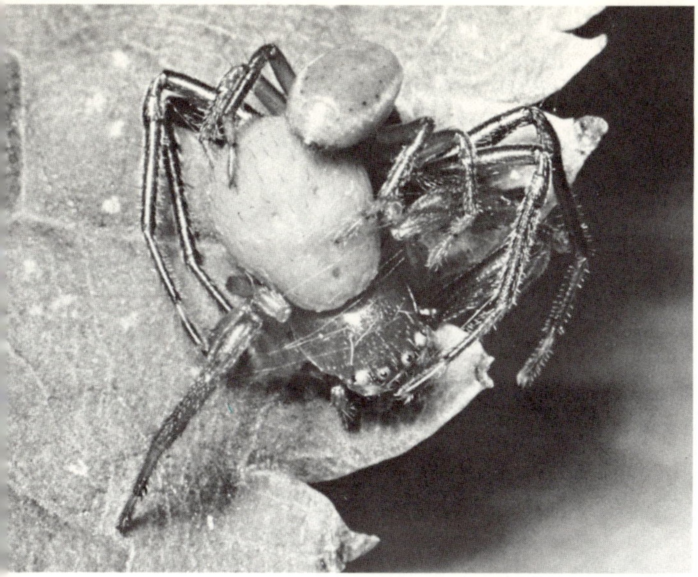

Fig. 86 The green species of *Diaea* mate in the normal thomisid fashion but the male first trusses up his mate. a) He inserts his left palp. b) Then he moves to the other side to insert his right palp. c) Mating having been completed the female is left to struggle free from her bonds.

no movement as the male approached and then halted at her side. After a short pause, he extended his front legs, touching first her legs and then her body while she retracted her legs and crouched. To all intents and purposes, the female appeared passive but to our surprise the male turned tail and ran. Reaching the edge of the leaf, he immediately dropped on his dragline hanging with legs outspread an inch or more below the surface. After hanging for a few seconds, he climbed rapidly up the thread, gathering it up as he went and on reaching the edge of the leaf dashed quickly to where the female still waited and then immediately climbed on her back. However, unlike *Sidymella* he kept moving across the carapace and legs of the female from side to side and this seemed to be what we were waiting to see. With the aid of a magnifying glass, we were able to confirm our suspicions – he was lashing her down securely to the surface of the leaf. Within a minute when she was firmly trussed he moved down to the end of her abdomen and crawled under one side in the same way we had observed with *Sidymella* (Figure 86, a, b, c). The mating went on from side to side for more than half an hour before the male departed; but there was no question of his now removing the restraining threads and his hapless mate was left to struggle free as best she could. We still do not know whether the species related to *D. albolimbata* share this behaviour but they probably do.

The egg laying habits of *Sidymella* do not seem to be known but those of *Diaea* are frequently observed. Once the eggs are ready to be laid, the female secretes herself away in a curled up leaf or in the half closed portions of the leaves of monocotyledonous plants such as Flax or *Astelia* and spins round herself a flimsy retreat. Some fifteen to twenty eggs are then laid, held loosely together with a cover of silken threads so thin that the eggs can be seen within. The mother spider now remains guarding them until they hatch some thirty days later.

CHAPTER TEN

The Six-Eyed Spiders

Most spiders have retained the primitive number of eight eyes but there are some families in which the anterior median pair was lost very early in their phylogeny so that all or most of the living species are now characterised by having only six eyes. With such an obvious characteristic to group them, it is convenient to consider them together. A few representatives of typically eight-eyed families have also lost one pair. There are a few of these in New Zealand but they are rare. In addition to the reduction in the number of eyes, this group, which in New Zealand includes three families, the Oonopidae, Dysderidae and Segestriidae, has also very simple reproductive structures. The palpal organ of the male is usually only a simple bulb as it is in the mygalomorphs and there is no epigynal structure above the epigastric furrow so that during mating the bulbs are inserted directly into the reproductive opening of the female.

THE OONOPIDAE

The six eyes of the oonopids are always placed close together and this characteristic, together with the fact that they are all small, makes them easy to identify. With the general run of these spiders, a body length of more than three millimetres would mean that for the family it was exceedingly large as the majority are much smaller than this, but in New Zealand we have a genus which includes the largest oonopids known. These comparative giants, which belong to the genus *Subantarctia*, may be all of five millimetres long and with their reddish brown carapace and legs and pale cream abdomen look very much like baby dysderids. Like all the oonopids, they are ground-dwelling hunting spiders which, while favouring forest conditions, are sometimes found near the margins if there is suitable cover. The first species to become

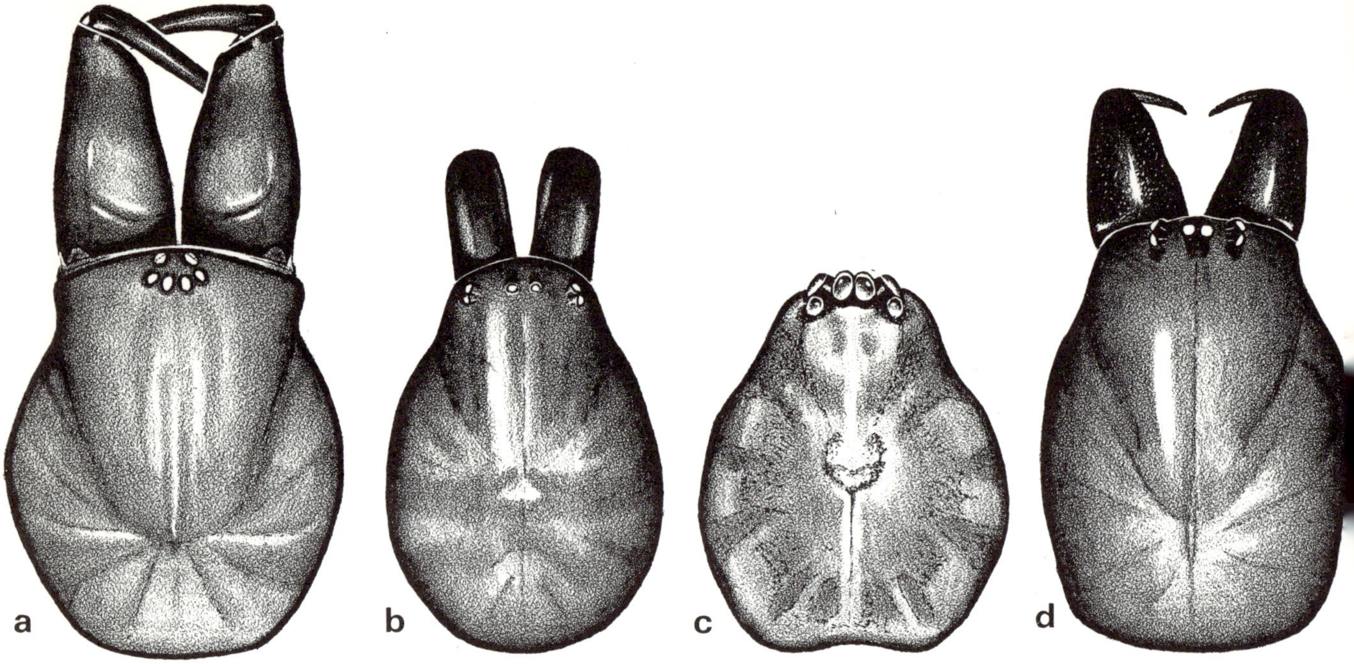

Fig. 87 The carapace and eyes of the six-eyed spiders. a) Oonopidae *Pounamua australis*. b) Dysderidae-*Dysdera crocata*. c) Segestriidae-*Periegops suteri*. d) Segestriidae-*Ariadna* sp.

known was found in Auckland Island in the sub-antarctic region but subsequently a number have been found in various parts of the South Island, usually during daylight, sheltering within a delicate tubular retreat beneath fallen logs on the ground.

The sixteen known species of *Ascuta* which range over all of New Zealand are characteristically inhabitants of damp leaf litter on the forest floor but some also thrive in the tussock grasslands. *Ascuta* consists of medium sized spiders rarely more than two and a half millimetres in length with the abdomen decorated with brownish or black chevrons on the dorsal surface and never protected with hard plates or scutes. It is the presence of scutes on the abdomen of *Duripelta* (Figure 88) which separates this genus from *Ascuta*. The abdominal scutes are usually restricted to the ventral surface in front of the epigastric groove in females but for some reason are much more extensive in males, where the plates sometimes extend back to the spinnerets and even to the upper surface.

The two further genera found in New Zealand are never more than one-and-a-half millimetres long. *Pounamua* includes six species with a dark purplish abdomen, brown carapace and legs while *Kapitia*, which is a very pale spider with a straw coloured carapace and cream abdomen, is known only from Kapiti Island.

THE DYSDERIDAE

The only representative of this family in New Zealand is the conspicuous cosmopolitan *Dysdera crocata*, some twelve millimetres long, which, with its bright reddish brown carapace and legs, contrasting with a pale cream abdomen, attracts immediate attention whenever it is seen. In addition to its striking appearance, the chelicerae are large and because they are directed forward give the spider a most ferocious mien which has little factual basis in fact as they rarely show any willingness to bite (Figure 89).

The recent work of Dr J.A.L. Cooke of Oxford University has increased our knowledge of the habits of this spider, which, in the wake of man, has achieved near worldwide distribution. One conclusion which was reached by Cooke is of some importance to gardeners because he found that this spider was almost alone in the animal kingdom in that it not only ate slaters but that these pests were its almost exclusive diet. On reading this information we checked through all the records of *Dysdera* in New Zealand – *Dysdera* incidentally seems to be recorded from most parts of the country – and we found that it did indeed occur in most places where slaters abound. The only question which still remains is why *Dysdera* is not more numerous in places like our garden where

Fig. 88 *Duripelta* (Oonopidae) lives among leaf litter on the forest floor.

Fig. 89 *Dysdera crocata* (Dysderidae) Male.

there is such a superabundance of delectable food for it. Surprisingly, the fact remains that it is never found in large numbers. It is a solitary spider which constructs a relatively capacious retreat near the ground, under stones or boards, leaving it at night for its leisurely chase. At the peak of summer, the bundle of eggs held together with only a few strands of silk is found in the retreat with the mother spider. Courtship with *Dysdera* does not seem a very auspicious occasion and mating takes place in an almost casual fashion. If the male should touch a female, she moves around until both spiders are facing, at the same time opens out her chelicerae and reaches forward with her front pair of legs. When contact is made the front legs of both spiders become entwined and they move towards each other with the fangs extended. The male taps the female with his front pair of legs while the second pair is used to stroke the underside of her abdomen as he rubs his palps along those of the female. The chelicerae of the male are held inside those of the female thus keeping them inactive. Having reached this stage, the male then inserts both of his palpal organs into the aperture at the genital furrow and the mating is completed.

One interesting aspect of this behaviour was examined in detail by Dr Cooke who concluded that recognition with these spiders was achieved mainly by touch and presumably by the exchange of some chemical substances, with little reliance on sight. He covered up the eyes of a number of the spiders he was studying and then watched them to find out if they could still mate successfully. He found that the loss of their eyesight had not the slightest effect on their ability to find each other and mate and it did not even prevent them catching their food so that some of them were still living a full life after some two years of induced blindness.

THE SEGESTRIIDAE

The segestriids are all moderately large spiders with their six eyes arranged in three pairs so that there is a definite space between each pair. Of the New Zealand genera only *Periegops* is rare, having so far been found only on Banks Peninsula in Canterbury. The single species, *Periegops suteri*, makes its home on the forest floor where it constructs a thin retreat under stones and fallen debris. The eye pairs are very widely separate and because it is not specialised for life in narrow tunnels the third pair of legs is directed behind in the normal way (Figure 90). The genera *Ariadna* and *Segestria* look very much alike and in fact it was only recently that C.L. Wilton was able to point out that some of the species which had previously been thought to belong to *Ariadna* did in fact belong to the related *Segestria*. Both are well fitted for life in narrow tunnels, with long tubular bodies and the three front pairs of legs directed forward. Fortunately, it is not necessary to delve deeply into the minor structural differences which separate these two genera because in New Zealand the colour patterns of the abdomen are enough to tell them apart.

Fig. 90 *Periegops suteri* (Segestriidae) Female.

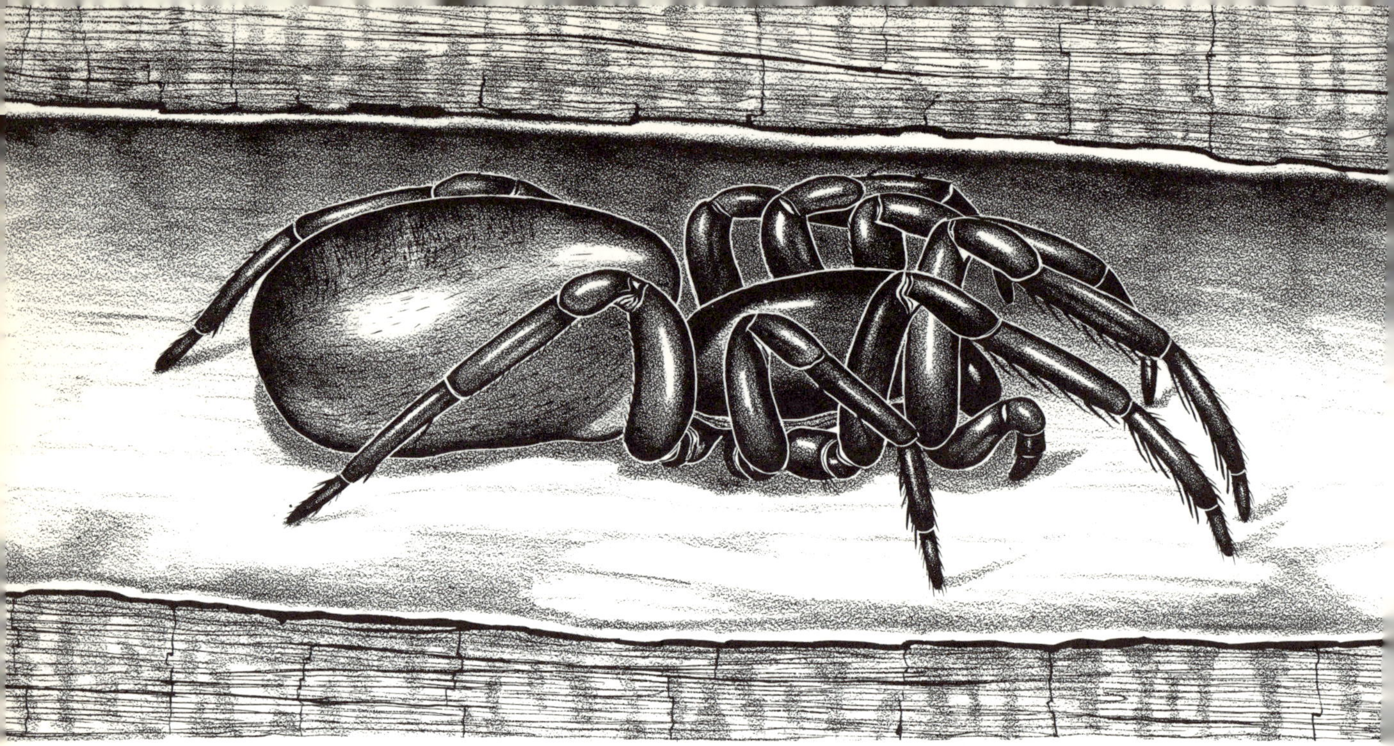

Fig. 91 *Ariadna* makes her home in burrows vacated by wood-boring insects.

Fig. 92 *Ariadna,* (Segestriidae) Female.

72. *Aranea pustulosa*. Female. Family Araneidae. Dunedin. Length 11 mm.

The Snare Builders

71. *Aranea pustulosa*. Female. Family Araneidae. Punakaiki. Length 11 mm.

73. Below left. *Aranea* sp. Female. Family Araneidae. Dunedin. Length 8 mm.

74. *Aranea* sp. Female. Monowai. Length 9 mm.

75. *Aranea viridata*. Female with egg-sac. Family Araneidae. Dunedin. Length 7 mm.

76. *Aranea viridata*. Female. Family Araneidae. Kaikoura. Length 7 mm.

77. *Aranea* sp. Female. Family Araneidae. Dunedin. Length 8 mm.

78. *Aranea atrihastula*. Female. Family Araneidae. Punakaiki. Length 6 mm.

79. *Aranea atrihastula.* Male. Family Araneidae. Punakaiki. Length 5 mm.

80. *Aranea* sp. Female. Family Araneidae. Bull Creek, Otago. 7 mm.

81. *Aranea crassa.* Female. Family Araneidae. Norsewood. Length 8 mm.

82. *Aranea crassa.* Female. Family Araneidae. Norsewood. Length 8 mm.

83. *Aranea pustulosa*. Female wrapping a fly. Family Araneidae. Dunedin.

84. *Arachnura feredayi*. Female. Family Araneidae. Stewart Island. Length 11 mm.

86. *Leucage dromedaria*. Female. Family Araneidae. Reinga. Length 7 mm.

85. *Argiope protensa*. Immature male. Family Araneidae. Mangareia. Length 2 mm.

Fig. 93 *Segestria,* (Segestriidae) Male.

Ariadna, the most widespread and common genus, has been recorded from the seashore to the mountains. Along the coastline, the narrow silken tunnels are occasionally found under stones but more often in the nooks and crannies on the rock faces above the high tide level. In open country, their retreats may be found beneath rocks but in the forest they prefer to use the holes vacated by wood-boring insects, sometimes in the smaller branches and twigs but also on the trunks of the trees. The tunnels are always well lined with tough silk which extends out to round off the mouth of the burrow so that the occupied burrows are easily seen. The abdomen of all but one of the species is brown or purplish in colour above and the pigment is usually darker down the middle to form a median streak. The one exception is *Ariadna septemcincta* which has a number of reddish brown chevrons down the dorsal surface of the abdomen (Figure 94). This species lives only in the forest where its burrow is often found in rotting logs lying on the forest floor and has also been recorded living in burrows in the ground.

Segestria in this country seems to live only in the forest where it also occupies old insect burrows in the trunks and branches of trees but there is no difficulty in identifying their homes, because, unlike any native species of *Ariadna*, a dozen or more stout threads radiate out from the mouth of the burrow. Perhaps this primitive snare demonstrates the way in which the first steps were taken by the primitive ground dwelling spiders towards the evolution of the multitude of aerial webs used by our present day spiders. *Segestria* rests at the mouth of the burrow with the front legs resting on some of the threads which are raised off the surface by small supporting pillars and thereby act as efficient trip lines. The spider itself is distinctive because the pale abdomen is adorned with a double row of conspicuous dark patches (Figure 93). The eggs are laid within the burrow by both groups of these spiders and when the young emerge they do not necessarily move away at once, for often a number of large but still immature babies will be found sharing the burrow with their mother.

Fig. 94 *Ariadna septemcincta* (Segestriidae).

CHAPTER ELEVEN

The Web Builders

PHOLCIDAE

Nearly all children in the North Island know what is meant when the daddy-long-legs spider is mentioned, but not so in the South Island because *Pholcus phalangioides*, while enjoying the protection of our homes, still seems to be susceptible to winter cold. Even with the increasing use of central heating, the future spread of this spider in New Zealand is by no means certain because, according to W.S. Bristowe who studied *Pholcus* in Great Britain, the dryness of centrally heated air inhibits the hatching of the eggs. The extended maze of fine threads which makes up the snare is usually placed high up near the ceiling with a definite preference for the corners. Somewhere amongst this network, the owner is normally found suspended upside down from the tips of its enormously long, slender legs, which when traced back meet at an elongate body (Figure 95). Few signs of life are seen until a small insect blunders into the web, and then as if with ten-league boots the spider strides over to the scene of the disturbance. But always it will stop well out of harm's way remaining at least a leg's distance from the struggling victim which it proceeds to truss up until further struggles are impossible. If too large an insect becomes entangled a very different reaction is observed, for the spider at once begins to shake its whole body so that all that can be seen is a blur as the entire web vibrates. It is not even necessary to wait for something large to be caught because a finger poked into the web will start it vibrating. It is probable that the purpose of this manoeuvre is to shake free those insects which would be too large for the spider to handle, because in addition to being rather fragile, *Pholcus* is one of the few spiders which does not have poison glands to assist in overcoming prey. Mating in *Pholcus* is of special interest, not because it is in any way spectacular but for the fact that the male inserts both of

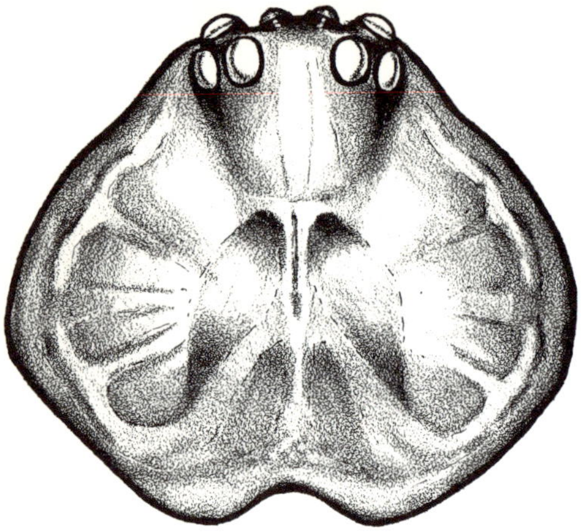

Fig. 95 Carapace of *Pholcus phalangioides* showing the typical triad grouping of the eyes.

his palpal organs simultaneously into the reproductive opening in the same way as many of the primitive six-eyed spiders such as *Dysdera*, and like this group, the female *Pholcus* does not have an epigynum.

It is usually agreed that the pholcids are among the most primitive of the web spiders. They guard their eggs in a most unusual way; after the bundle has been laid, they are held together with a few threads of silk and carried about in the chelicerae of the female until they hatch. The egg-bearing females are common in New Zealand from about Christmas but continue to carry successive batches right up till early Autumn. The young leave the mother as they hatch and hang suspended, often like a line of drying clothes, from some of the threads of their mother's web. Each batch of eggs takes about three weeks to hatch and according to the studies made by the French arachnologist, Pierre Bonnet, must pass through five moults before they reach maturity. Bonnet found that at the height of the season this might take about twelve weeks but as the weather became colder and food less abundant growth would become slower and those spiders which did not pass through their fifth moult before the autumn set in did not reach maturity until the winter had passed. No one has checked to see if these observations are true for the spiders in this country.

Although the only pholcid which most people will see is the common daddy-long-legs in and around the house, at least two more smaller species with the typical long and slender legs but a shorter, almost oval, abdomen have also been found. One of these species, which is often common under rubbish lying about in gardens and backyards in the North Island, has an attractive, greenish abdomen broken up by two pairs of white patches while the other, found in the South Island, is a uniform brown. Because these pholcids have never been found away from their urban haunts they have probably been introduced and eventual identification may well provide a clue as to their place of origin.

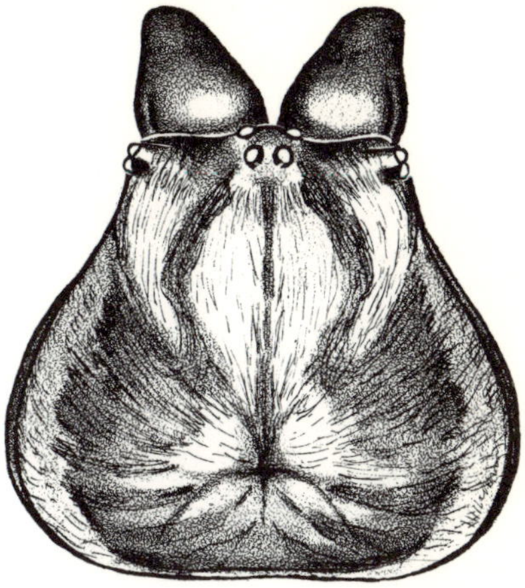

Fig. 97 Carapace of *Aranea pustulosa.*

Fig. 96 The lightly bound egg bundle of *Pholcus phalangioides* is held by the chelicerae and palps.

ARANEIDAE

The conspicuous cartwheel webs spun by the members of this family ensure that they are well-known and even as they become accepted as part of the landscape they cannot but at times raise a sense of wonder that such seeming perfection in symmetry and structure can be achieved by such lowly animals. Is this not the outcome of some form of intelligent reasoning or at least the end result of some period of learning passed down from the parent spider? This can be disproved, however, as the parent spider is often dead before the spiderling is ready to construct its first snare and if we remove the egg-sac and isolate the spiderlings as they emerge, they are still capable of passing smoothly through all the steps required to produce a perfect and characteristic snare. We actually find, on a close study of the steps taken to build this structure, that the spider is restricted to a very narrow path where one type of behaviour must precede the next. If the earlier stages are partly destroyed, it is quite unable to work out that it would be useless to continue without repeating them but carries on regardless even though the final result is a travesty of the structure the spider set out to create. While the stimuli triggering the chain of actions leading to the finished web are by no means fully understood, by observing the spider in action it becomes clear that the striking end result is achieved by means of a series of simple actions.

Most orbweb spiders construct their webs after dark but many will begin during daylight hours if the weather is dull and overcast; this is the best time to observe them (Figure 98). First the spider bridges a gap to provide the transverse line from which the web is hung. This bridge thread is usually laid down in one of two ways. The spider may attach one end of the thread to the twig where it rests and then walk down and around while holding the thread free with one of the fourth pair of legs until it reaches another twig a suitable distance away. If, however, there is no easy route to use to accomplish the task, the wind is used to carry a floating thread across the gap until it becomes firmly anchored. Once the gap is bridged a thicker line is transported across and the original line removed, so that this new line remains as the basis for the web to come.

The next series of actions is concerned with establishing the outer boundary threads and the spokes which will radiate out from the centre of the finished web. The upper sector is constructed by the spider moving across the bridge line while it carries a further line held free with its legs. After this line is fixed, it moves back to the centre which it seems to judge by the even tension of the thread on each side, and after attaching a further thread at this central point drops down until it reaches another twig below. The vertical thread on which it has dropped is then pulled tight and firmly fixed. The pull of this thread stretches the transverse line to which it was earlier fixed so that now the two threads take up a Y-shape with the future centre of the web where the three arms meet. Next the spider will run another line as it moves along one of the upper arms and then after attaching yet another thread to the centre of this, carries this thread up to the bridge line to be fixed and pulled tight, so that we now have another radial in place and a further transverse line below the bridge line which will form the upper limits of the orb itself. The spider continues to apply further radials along with marginal threads to the lower sectors in the same way until, supported by a few widely spaced radials, the outline of the web takes shape.

The next stage is to fill in further radial threads until the full complement is attached. To do this the spider moves down an existing radial to the hub and, after attaching a further thread, carries it out to the boundary line then moves along this line a suitable distance before attaching the thread it is carrying to form the next radial. By continually moving from the hub of the web to the outer margin the remaining gaps in the array of radials are quickly filled in until the whole area is covered with evenly spaced rays. At intervals, the spider may be seen revolving at the centre hub and laying down the threads which will hold the radials in place until it continues on to the next stage when the spiral scaffold will be laid. When this stage has been reached, the behaviour abruptly changes and the spider begins to circle out from the centre, attaching a widely spaced spiral thread of dry silk as it goes; this will both keep the radials in place as the sticky spiral is laid later and provide a footing as it moves around. The final sticky spiral is begun from the outer margin with the spider rolling up the scaffold thread as it goes. To lay this permanent spiral, the spider moves up and down the radials and crosses between them on the scaffold thread carrying the line of silk, which is now coated with a sticky substance, as it goes. The sticky spirals are laid much more closely than the earlier scaffold spiral and the web gradually assumes its familiar form. The sticky material is at first evenly coated over the inner thread but as the spider attaches each sector across the gap between the radials it stretches the line so that the surface tension of the fluid is broken, then the sticky coating gathers like a string of beads fairly evenly spaced along its length. The spider ceases to lay down the sticky spiral when it is still some distance from the hub so that there is always a clear area near the centre where it will normally rest.

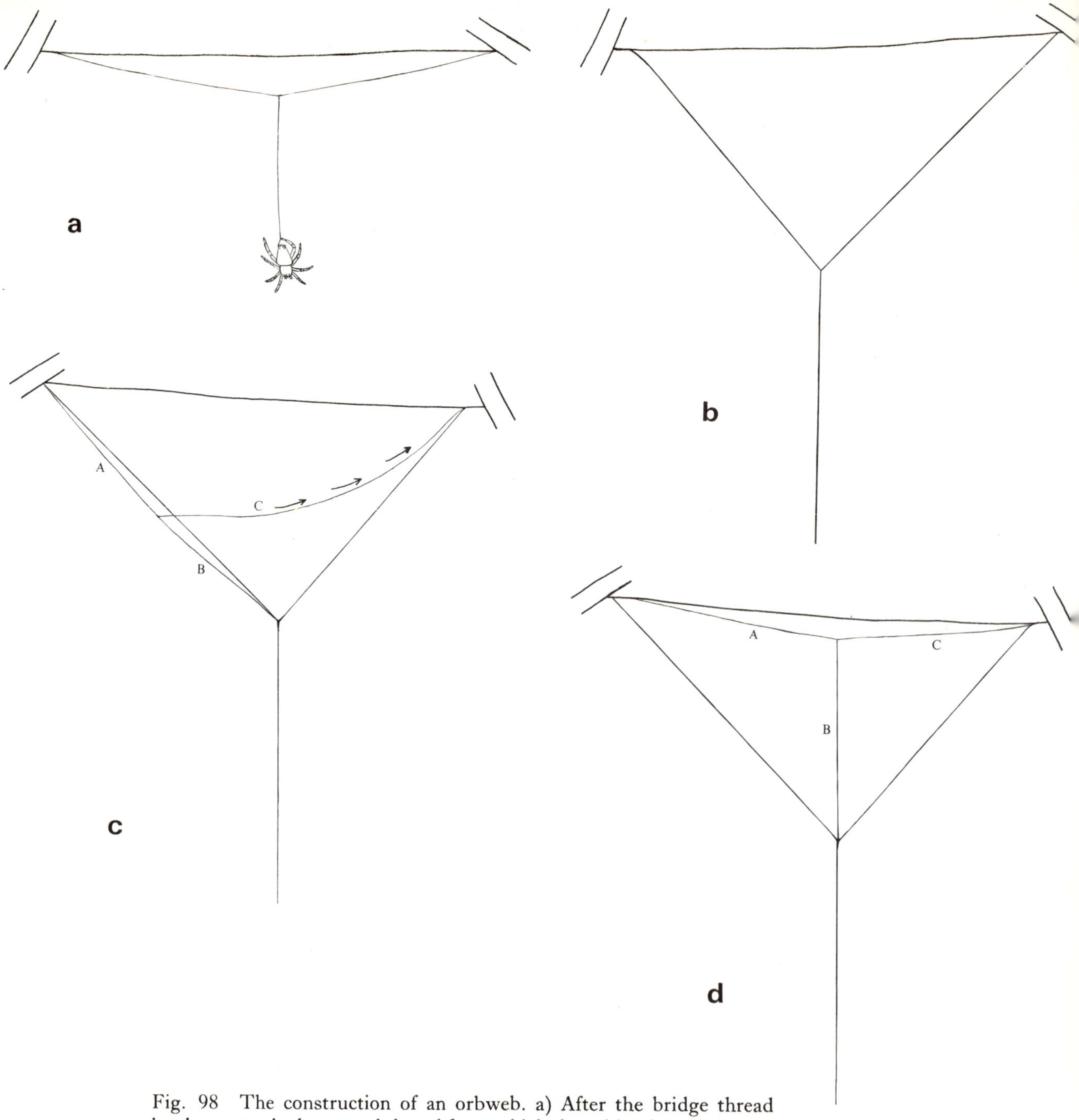

Fig. 98 The construction of an orbweb. a) After the bridge thread has been attached a second thread from which the spider drops is strung out b) The vertical thread is attached and pulled tight so that the first three radials are formed. c) A further thread is laid along one of the upper radials and connected to the opposite corner by the thread C. d) Thread C is pulled tight and one section is completed. e) Other sections are completed in the same way. The inner radials are then filled in one by one by the spider carrying out a thread from the hub and attaching it to the outer margin thread. f) The scaffold thread of dry silk is laid in a wide spiral from the centre. g) Using the scaffold thread and the radials for a footing the permanent sticky spirals are laid from the outside in, gathering up the scaffold thread as it goes. h) As each section of the sticky spiral is attached it is stretched so that the uniform sticky coating is broken up into beads.

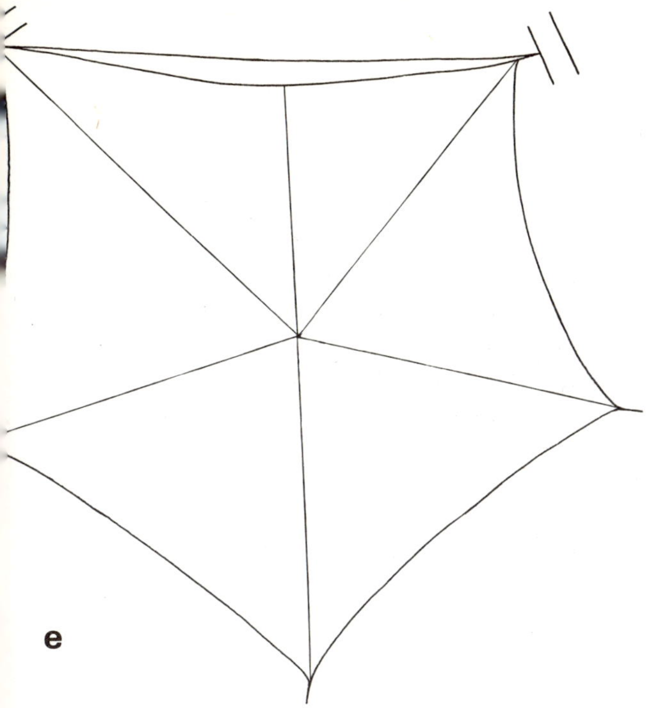

e

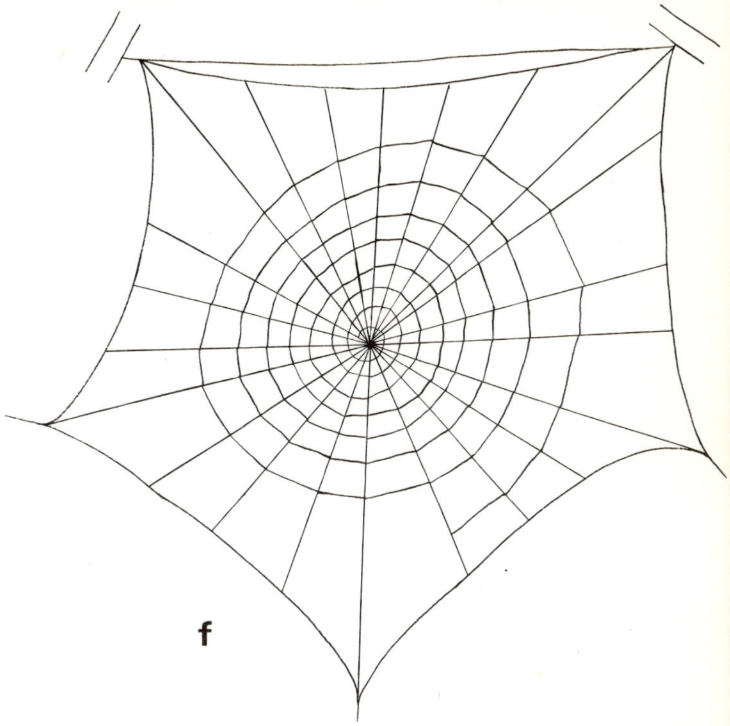

f

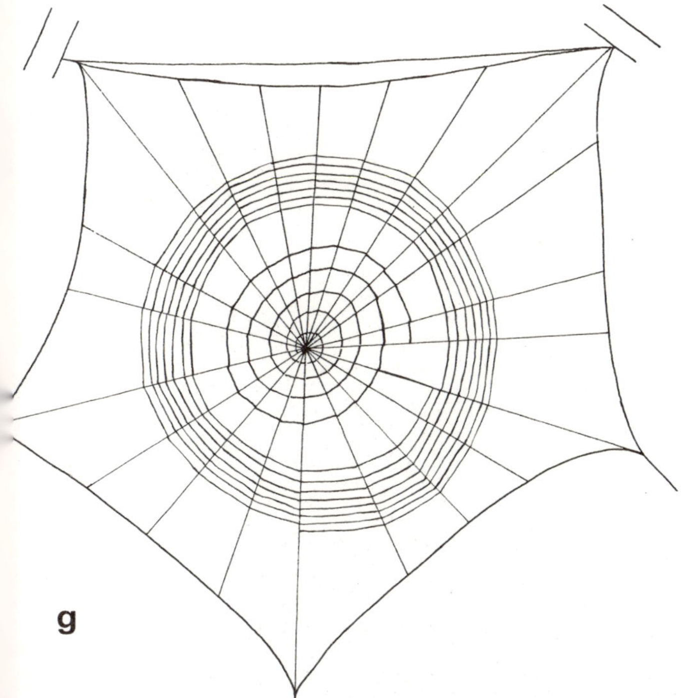

g

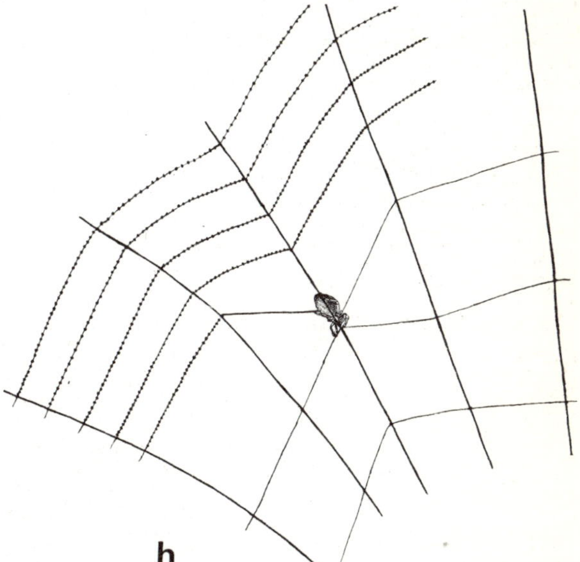

h

The web is now ready for use but many spiders carry out further modifications. Some will clear out the threads from the hub to leave it clear, while others will spin conspicuous zig-zag sheets of silk out from the hub to form the stabilimentum, the purpose of which is variously claimed to provide additional stability to the web or to make the spider difficult for predators to see. It is not easy to describe the steps the spider takes without making it sound extremely complicated but half an hour spent watching one at work will show the whole process as a smooth progression where one step logically leads to the next until the web is complete.

There are two questions which inevitably come to mind if these spiders are closely watched. The first is how does the silk from the scaffold line disappear after the spider takes it up? The answer is that it is usually eaten, not as dry silk but by first breaking it down into a liquid with the digestive enzymes which ooze from its mouth; in fact the same thing happens later to the silk which the spider uses to wrap

Fig. 99 Partly constructed orbweb. The widely spaced spiral is the scaffold thread, replaced on the outer margin by the more closely spaced permanent sticky spirals.

up its prey. The second query is why the spider does not become trapped in the sticky threads of its own web and here the answer is that it is protected by a coating of oil which is continually renewed from glands also associated with the mouth. An interesting experiment may be carried out by dipping the legs of one of these spiders in a fat solvent, such as ether, and then putting it back on the web. At once you can see the difficulty it experiences until the coating is renewed.

After the web is complete, the spider will take up its position at the hub and cling motionless head downwards from the claws at the ends of its legs until the movements of a trapped insect bring it rapidly to life. Its first action is usually to tweak at the threads which will often set the insect into movement again but probably the main purpose is to enable the spider to select the threads which have the greatest tension to use as a path leading unerringly to its prey. On reaching the trapped insect, it first touches it with the legs or palps and, if the prey is not too big, bites it and then twirls it round and round, wrapping the silk around the victim like thread on a bobbin. A further quick bite may now be administered before the wrapped-up bundle is cut free from the web and transported to the hub where it will be eaten.

After a good night's catch, the web will be badly torn but few of these spiders are able to make effective repairs. Some spiders will use the web for some days with the gaps bridged by old draglines but many will tear down the mutilated structure and construct a new web on the framework of the earlier one. We found that the beautiful green *Aranea viridata* which we had photographed in our garden belonged to this house-proud group as it regularly rebuilt its webs every day for more than the week it took us to get the photographs we wanted. During the daytime, many of the webs are empty but this does not mean that the owners have gone. A strong thread inevitably leads away from the hub of the web to some secure hideout nearby where the spider remains with the claws of a leg hooked on to the cable which then transmits vibrations to her from prey caught in the mesh. No-one seems to have recorded the mating behaviour of any of our species of orbweb spiders; all we can do is to outline what we might expect to find and hope that some of our readers might be prompted to fill in this gap. As with most spiders, the males are smaller than the females and in some, such as *Arachnura* and *Celaenia*, they are truly midgets. It is probable that some of these minute males are able to approach a mate almost without her knowledge but for most of the males there is first an urgent need for an approach to be accepted before he becomes the subject of her next meal. This preliminary recognition is normally achieved by a series of signals which the male transmits by tweaking at the threads of the web as he slowly approaches his prospective mate. At first he may be repulsed and then rapidly retreats or drops out of harm's way on a thread, but the drive to mate is strong and he keeps repeating his attempts until the absence of any further aggressive reactions encourages him to come quite near. However his troubles are not all over even then because, after spinning a special mating thread, he will now need to coax the female with a series of jerks and tugs to take her place on this thread where she will hang upside down. If he succeeds this far, most of his problems are over. As the female hangs motionless upside down on the thread to which he has enticed her he will move up and, by clasping her with his legs, will insert his palp while they both hang downwards.

The orbweb spiders most commonly seen belong to the huge world-wide genus *Aranea* which in general can be recognised by the broadly triangular shape of the abdomen when it is seen from above. As may well be imagined, these common and easily collected spiders attracted the attentions of our earlier spider students in the late nineteenth century who named some sixty species from New Zealand. Unfortunately, the preoccupation at that time was with the colour patterns which may vary from specimen to specimen; at present we are not sure to which species the names belong, although from collections in the various museums it is clear that almost a hundred species occur. The commonest species

Fig. 100 *Aranea pustulosa* (Araneidae) female.

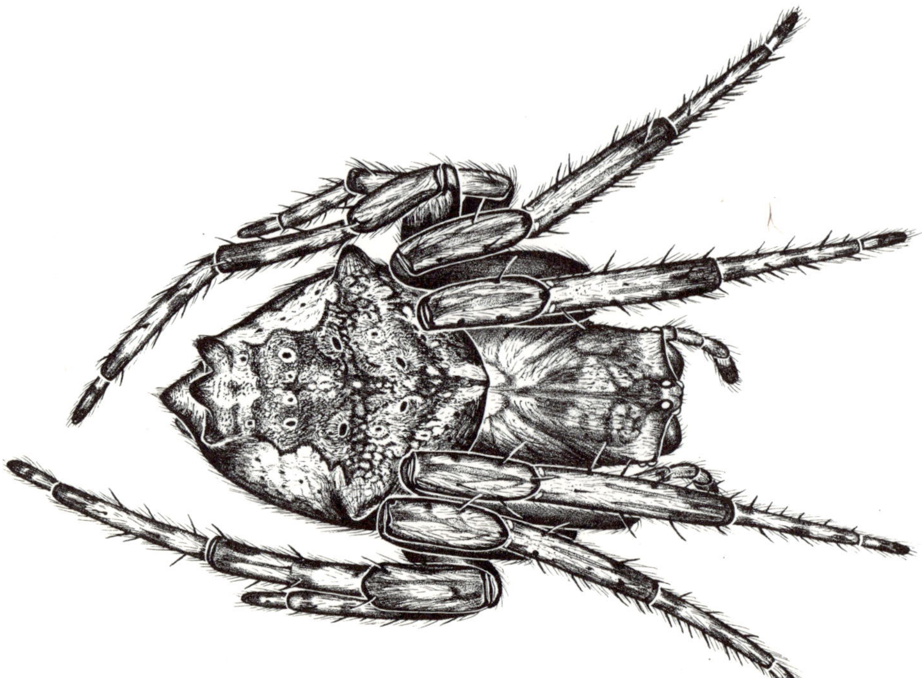

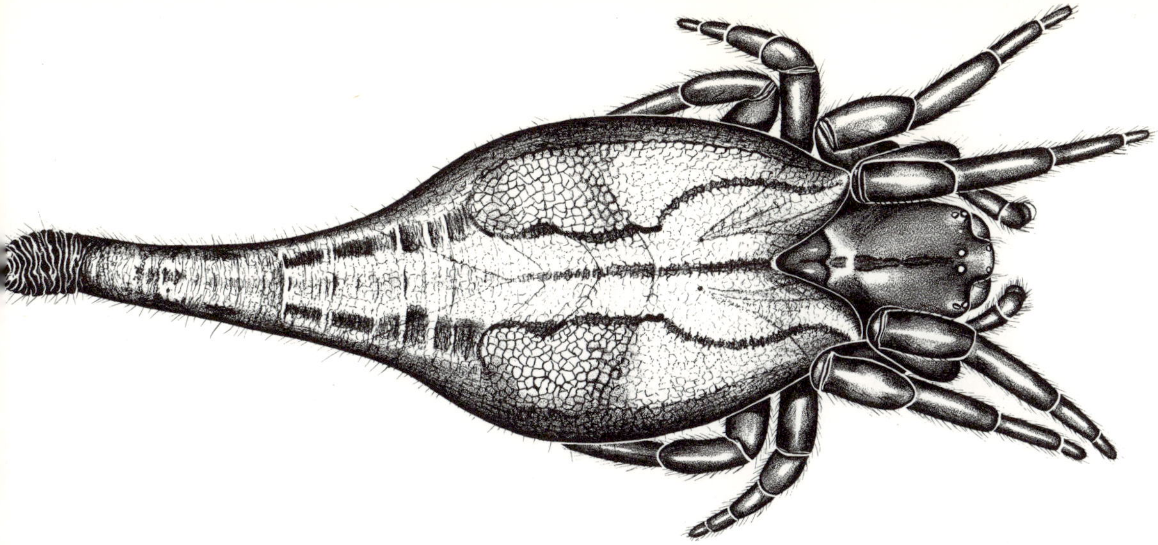

Fig. 101 *Arachnura feredayi* (Araneidae) female.

throughout New Zealand is *Aranea pustulosa*, which we share with Australia (Figure 100). It is not certain whether this spider found its own way across the Tasman Sea or was brought in during colonization but it has certainly been in this country for a long time. It may assume an unending series of colour patterns ranging from pure white on top of the abdomen through greys to black and all shades of yellow and brown. Fortunately, it can always be recognised by the five small knobs at the tip of the abdomen which are arranged in a transverse row of three,

Fig. 102 *Arachnura feredayi* climbing up her dragline with the bundle of drawn up silk.

followed behind by two in a longitudinal row. It is usually the commonest spider in the garden and now occupies any moderately open space although it still does not seem able to penetrate far within the forest. The egg-sac embedded in loose, wiry silk, at first brown, but later turning to an olive green, is attached to a twig near the orbweb but remains connected to the web by a thread, where the mother spider will often cling.

Another orbweb spider commonly found on hedges or amongst the gorse and scrub is a greyish brown one with a prominent white line across the front of the abdomen, called *Aranea crassa*. The white band is usually the most distinctive feature but sometimes this is broken into a pair of white spots or another white band down the centre to form a T-shaped mark with the front band. A number of pretty orb-weavers which are pale green on top of the abdomen are found near the margins of forests and also in gardens. The species illustrated in colour, *Aranea viridata* which is common in Dunedin, constructs a conspicuous stabilamentum in its web. The egg-sac is similar to that of *pustulosa* and as with this species is found bound to a twig near the web. Within the forest are found mottled green and brown species such as *Aranea atrihastula* which leave their webs during daylight to rest on the trunks of adjacent trees where their colouring merges closely with the surrounding cover of moss and lichens.

The tailed spider *Arachnura feredayi* is an extraordinary spider (Figures 101, 102). The elongated yellow-brown female has a three-quarter inch long body which is more than one-third tail. The tail is slender and can actually be wriggled but it is difficult to imagine any particular use for it as the spider rests motionless at the hub of her web, nor can the tail perform any role when she captures her prey because the spinnerets are not at the tip of the tail but half-way beneath her body. The web, usually strung close to the ground, is unusual in a number of ways. The upper sectors are never filled in with the sticky spiral so that the web appears unfinished and is always surrounded with irregular threads. The egg-sacs are pale brown but, instead of being hidden away, are placed in full view in a string along the open upper sector of the orb with the lowest egg-sac at the hub where the spider rests head downwards. As many as six or seven of these sacs may be found all bound together and each sac may contain up to fifty eggs. The more recent egg-sacs seem to be the ones near the centre of the web and this puzzled us until we noticed an account by J.F. Jones of the habits of the closely related *Arachnura higginsi* from Tasmania. He observed that the spider always constructed the egg-sac at the centre of the web then took her place below it making this position the centre for the new web she constructed next day after the old web had been demolished. By its carrying on in this way, it can now be seen that after six or seven egg-sacs have been laid in a line, the position of each new web will become much lower than the original web. The lower end of the string of egg-sacs where the spider rests will thus enclose the more recently laid eggs. The first encounter with the male of this species engenders disbelief that a spider little more than one millimetre long and perhaps 1/150th of the bulk of its mate could effectively consummate such a union.

Quite apart from its size, the male looks very different at first because only a slight pointing of the back of the hard abdomen shows where the tail should be. The mating of our species has not been seen and so we will need to refer again to Mr Jones' observations on the Tasmanian *A. higginsi* to anticipate the procedure. He found the minute males which lurked in the irregular threads at the outskirts of the orbs approached the females in the usual way, vibrating the radial thread during the approach. It would seem from these observations that the special mating thread was constructed near the hub where the female rests and after enticing her on to it mating takes place in a typical way.

Arachnura higginsi may be found in large communities where each spider has its own orb but these are often interconnected. This does not seem to occur in the New Zealand species although it is commonly found in the next orbweb weaver we will mention. The female *Cyclosa trilobata* is easily recognised by the stout but short tail, flanked on each side by a prominent swelling (Figure 103). The general colour of these spiders is glossy black with silvery patches down the upper surface of the abdomen but some specimens also have reddish markings. The male, less than a quarter-inch in length, is only half the size of the female and the three lobes at the end of the abdomen are smaller. Although a single isolated web is occasionally found dozens of webs are frequently grouped together, usually in some shady spot and never more than a few feet from the ground. The webs are typical orbs but a stabilamentum is woven across each side of the central orb in which debris is incorporated and the spider rests at the hub so that its body is in line with this structure, and is then very difficult to see. Unlike *Arachnura*, the egg-sacs are not placed in the snare but deposited on a nearby twig. Each sac is in the form of a flattened lobe of fine brown silk and as a number are usually placed side by side the end result may often be mistaken for the seed pod of the Kowhai.

Another tailed spider, most commonly seen in the North Island is *Argiope protensa* which is to be found at the hub of its large orbweb

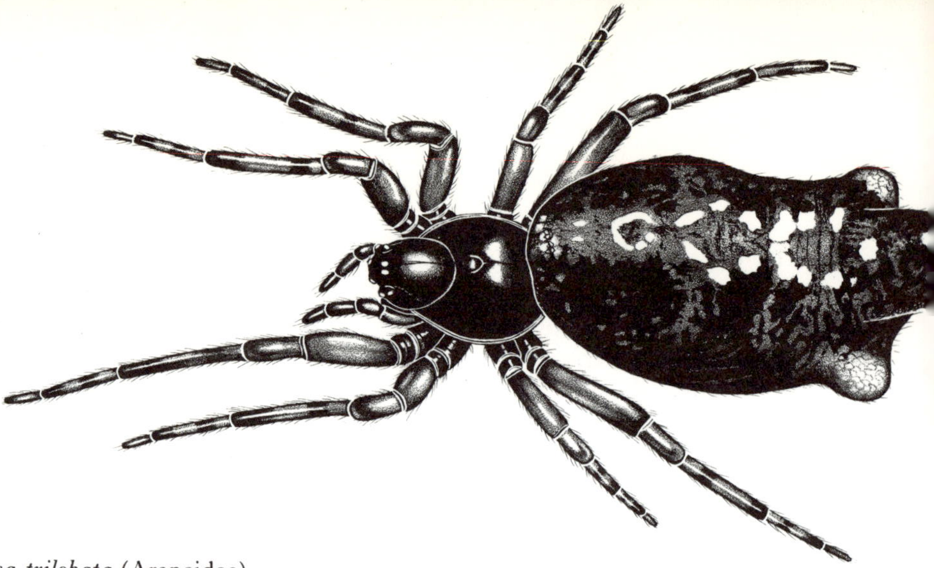

Fig. 103 *Cyclosa trilobata* (Araneidae)

during summer. Its body may be three-quarters-of-an-inch in length and the tail is short and stout. The carapace gleams with a thick covering of silvery hairs while the silver abdomen is suffused with yellow along the median surface and sides. Because the posterior portion of the abdomen is a tail, the spinnerets are on the under side. The yellow shading on the abdomen becomes more pronounced as the spider reaches maturity so that young specimens may appear almost entirely silver. The webs are often built at an angle to the ground and many have some form of stabilamentum. The smooth greenish coloured egg-sacs in the form of a high dome are attached to a twig somewhere close to the web, (Figure 104). The male is very much smaller than the female and, although the mating of *protensa* has not been observed, it probably follows the pattern of its relatives overseas where the female binds the male with silk from her spinnerets as he inserts his palps; but it would appear from these accounts that this action does not necessarily indicate any sinister designs on the female's part as, at least in the first mating, the male may eventually manage to escape.

The most conspicuous orbwebs found in the forest may be three feet wide and are usually strung horizontally over water. The long-legged spiders associated with these webs belong to the sub-family Metinae and are usually placed in *Orsinome* (Figure 105), and many different species can be found in different parts of the country. Apart from size, these spiders all look alike and are characterized by the extreme length of the front two pairs of legs which are stretched in front where they rest in their web. The leaf-like pattern on the top of the abdomen is always present and the general colour is brown. The most extraordinary orbweb spiders we have in New Zealand are the Bird-dropping spiders of the genus *Celaenia* (Figures 106, 107). The carapace is

87. *Orsinome* sp. Male. Family Araneidae. Mangareia. Length 8 mm.

88. *Celaenia* sp. Female. Family Araneidae. Dunedin. Length 4 mm.

89. *Celaenia* sp. Female with egg-sacs. Family Araneidae. Wairoa. Length 4 mm.

90. *Tetragnatha* sp. Female. Family Araneidae. Dunstan. Length 8 mm.

91. *Tetragnatha* sp. Male. Family Araneidae. Buckland's Crossing, Otago. Length 6 mm.

92. Forest *Tetragnatha*. Male. Family Araneidae. Peel Forest. Length 4 mm.

93. *Mimetus* sp. Male. Family Mimetidae. Dunedin. Length 4 mm.

94. *Achaearanea veruculata*. Female with egg-sac and young. Family Theridiidae. Christchurch. Length 5 mm.

95. Top left. *Achaearanea veruculata*. Female. Family Theridiidae. Punakaiki. Length 5 mm.
96. Top right. *Achaearanea* sp. Female. Family Theridiidae. Bull Creek, Otago. Length 4 mm.
97. Left. *Steatoda* sp. Female. Family Theridiidae. Napier. Length 7 mm.

98. Katipo, *Latrodectus katipo*. Female. Family Theridiidae. Length 6 mm.

100. Female katipo *(Latrodectus katipo)* guarding her egg-sacs.

99. Katipo *(Latrodectus katipo)* mating. Family Theridiidae. Christchurch.

101–103. A female Katipo grows up. 101. The spiderling in early February immediately after emerging from the egg-sac. Length 1.5 mm.
102. The same spiderling in June. Length 2 mm.
103. In August it is now only three moults from maturity. Length 3.5 mm.

104. The male katipo retains the youthful colouring. Length 4 mm.

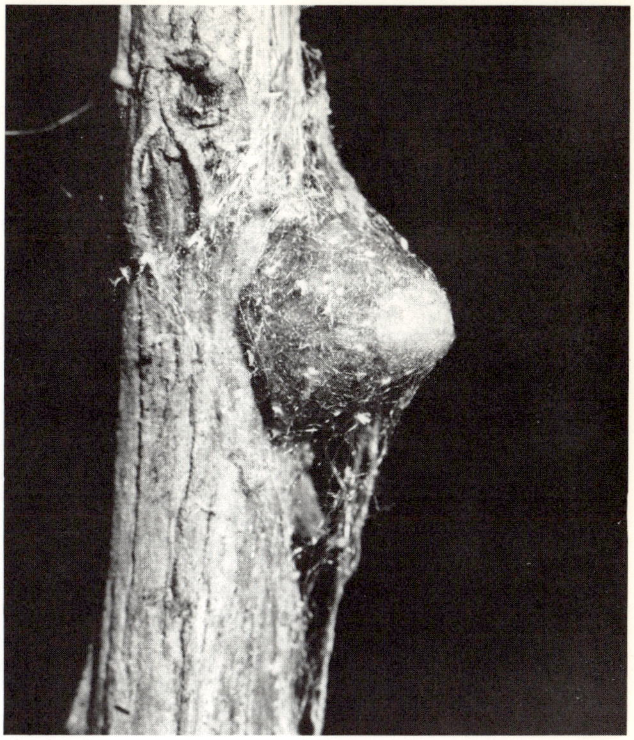

Fig. 104 Egg-sac of *Argiope protensa*

always drawn into a narrow peak in front where the eyes are grouped together but it is the abdomen which always seems so grotesque. Unlike most spiders, this is drawn out at the sides so that it is much wider than long. The dorsal surface has peculiar bumps and projections so that when the spider is at rest with the legs drawn up to the side it is difficult to pick it out from the debris or bark where it lies. Some are dirty white and consequently it can be seen how the name bird-dropping spider was derived. Quite apart from their unusual appearance, these spiders merit another claim to fame because they do not construct an orbweb at all; in fact, they seem to have abandoned completely the use of silk for the capture of their prey. They are usually found clinging to the leaves of low shrubs or perhaps hanging upside down an inch below a leaf,

Fig. 105 *Orsinome* sp. (Araneidae) female.

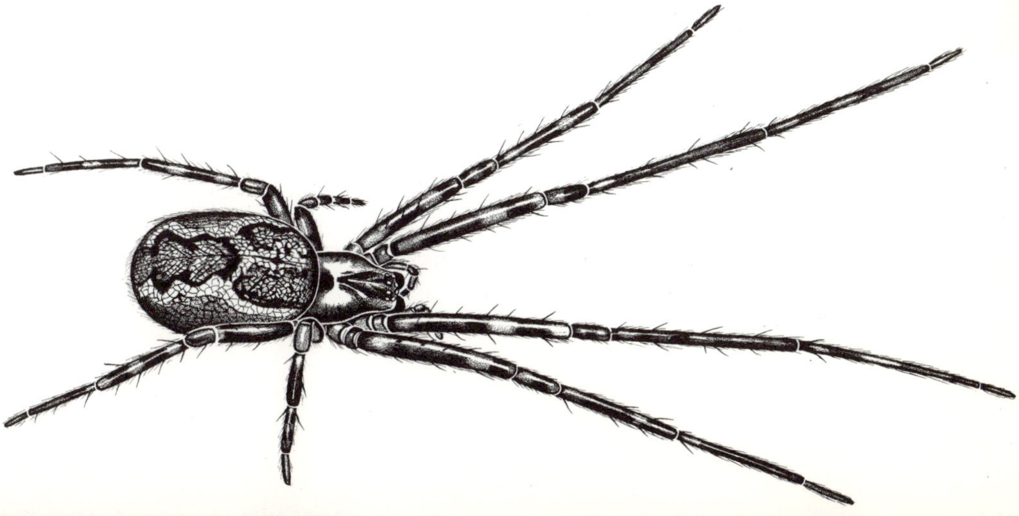

clinging with the front pair of legs to a thread still attached to the spinnerets.

Similar species are found in Australia where they are called Orchard spiders and sometimes the Death-head spider because of their resemblance to a skull. It is to these spiders we have to turn to get some possible clues to the habits of our own species. For many years, it was not known how these spiders captured their prey but this has become clearer since some interesting observations were made in Australia. It seems certain that the normal food consists of moths which the spider grasps with the heavily spined front two pairs of legs as it hangs down from a thread of silk. That this is so of course immediately raises another question. How often would a moth pass close enough for the spider to make a successful capture? With this point in mind, a close examination was made of the moths *Celaenia* caught and it became evident that they were all males. Furthermore, after a careful watch, it was found that the moths flew up to the dangling spider as if they were attracted to light or, much more likely, as if they were approaching a female moth of their own kind. While it is still a matter of speculation, it is now thought that these spiders exude a scent which has the same effect on the male moths as the mating scents produced by the female moths to attract a mate. If this is so, it would account for the spider's success in obtaining its food and also the fact that it catches only male moths.

We were most anxious to see if our own species acted in the same way but, unfortunately, these spiders are rare in the South Island and it was only a short time ago when we obtained a supply of females and egg-sacs from Mr Bernard Teague of Wairoa. While these spiders behaved in the same way as the Australian species and would take moths from tweezers when we offered them, we have not seen them catch a moth for themselves. The males are always extremely small and we wondered whether, as with other related spiders, they might be nearly mature by the time they left the egg-sac. The round egg-sacs are each a quarter-inch wide covered with a dense dark-brown silk with four or five knobs around the sides (Figure 106). Each egg-sac has a stout stalk but, as the female continues to construct her full quota of seven or eight sacs, they are held in a bunch with a light envelope of threads rather than by the stalks. After waiting for some weeks for the spiderlings to emerge, we decided to see if we could expedite matters by regularly spraying the sacs with water, and, sure enough, within two days the spiderlings began to emerge. At intervals of a week to ten days, further sacs released their young which all emerged from a neat round hole on the upper

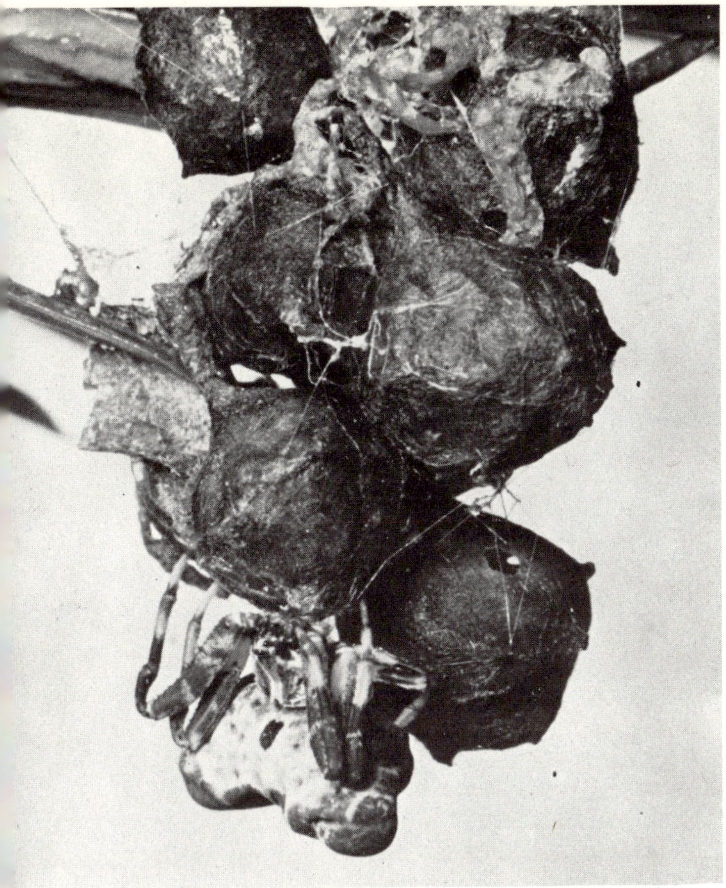

Fig. 106 *Celaenia* sp. (Araneidae) from Wairoa a) Clinging to bundle of egg-sacs. b) Suspended from a silken thread she patiently waits for prey.

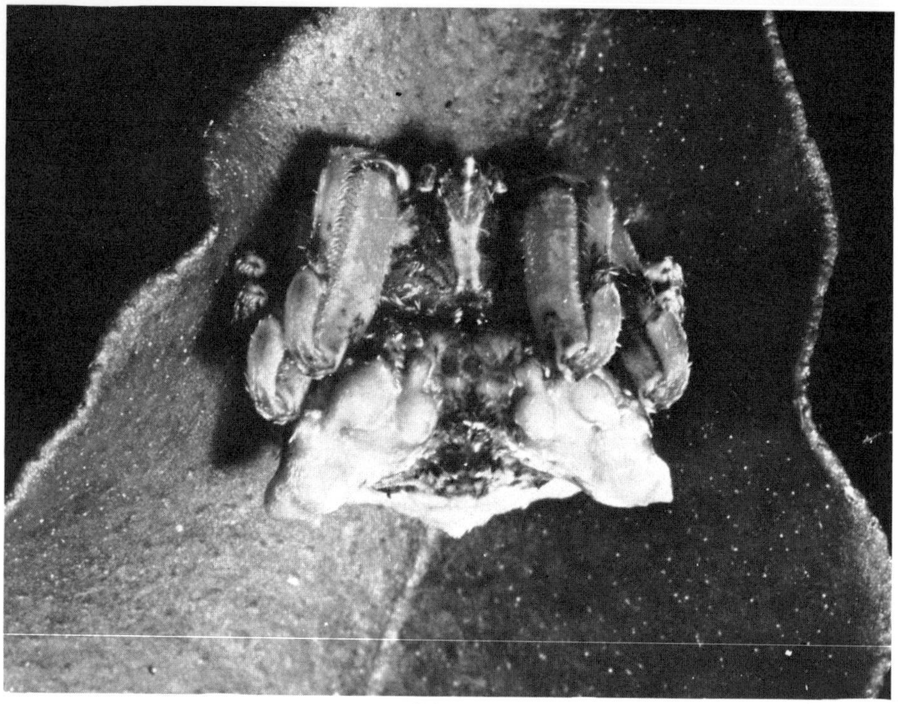

Fig. 107 *Celaenia* from Dunedin.

surface of the cocoon. Our first hopes were dashed when we saw their size because they were still much smaller than a fully grown male; a close examination soon showed that these pale little spiders with strong bristles on the abdomen were all the same and it would evidently be some moults before the males could be distinguished. The minute spiderlings moved to and fro over the twig laying down their draglines as they went until all the leaves were clothed with fine silk.

We next wanted to know whether the young *Celaenia* would construct orbwebs to trap some food until such time as they were big enough to catch moths. We do not know this yet as the spiderlings do not seem interested in food but many have begun to suspend themselves on a thread of silk in the same manner as their mother. Their expectations in respect of food remain a puzzle because even the smallest fly would seem to be beyond their capabilities, let alone a moth.

Leucage dromedaria, about eleven millimetres long with greenish legs, a pale carapace and a conspicuous silver abdomen with a prominent hump on the front is a truly striking spider but in fact is rather difficult to see. The orbweb is typical but strung horizontally amongst the undergrowth of the more open type of forest, with a maze of threads below. The spider rests on the underside of the orb so that the ventral surface, which is dark, merges with the dark background of the forest floor as you look down upon it. The webs are usually constructed near water where a number may be found together to form a small colony.

Tetragnatha is often placed in a separate family, the Tetragnathidae but here we will discuss these spiders along with the rest of the orbweb weavers (Figure 108). These are the spiders which construct most of the orbwebs with widely spaced spiral threads found near water, particularly open swamps. The long, thin body, the elongate legs and the strong heavily-toothed chelicerae separate them from other spider-like forms with which they might be confused.

Another characteristic of the tetragnathids is the absence of an epigynum in the female and the simple form of the palpal bulb in the male. They occupy their webs mainly during the evening but may be seen during the day if the weather is dull. When at rest, the front two pairs of legs are stretched forward and the two hind pairs behind so that the spider looks just like a twig caught up in the threads. Most of the species are drably coloured in various shades of yellow and brown but a group of small species which live only in the forest may be green and red. These attractive little forest tetragnathids string their webs horizontally across the low foliage and do not seem to be particularly concerned with living near water. The habits of neither our own species nor the related Australian ones have been studied much and so we can record little of their life histories at present but it is likely that the mating behaviour follows closely that of species of *Tetragnatha* from other countries. Like the *Tetragnatha* described from overseas, the chelicerae of the males of our species possess exceptionally strong teeth near the tip which indicates that their habits are the same. It has been found that there is no obvious courtship as the male approaches his mate. As their front legs touch, they move closer so that both spiders have these legs pointing more to the side. As the chelicerae touch, the fangs of the male clasp the chelicerae of the female, firmly wedging them by means of these specialized uppermost spines. This first approach is made while both spiders hang upside down but, once the chelicerae are locked, both spiders drop down on the threads so that they are hanging vertically, venter to venter while the male applies his palp.

Fig. 108 *Tetragnatha* sp. female.

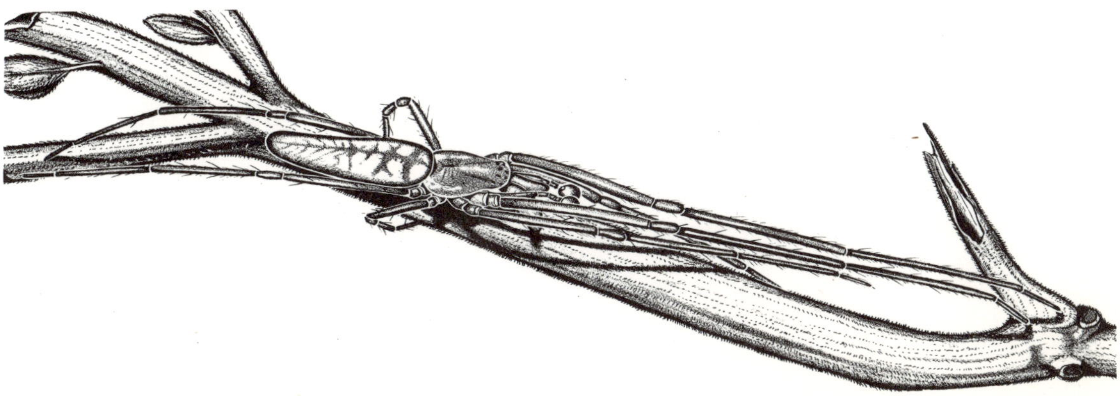

MIMETIDAE

Several very pretty spiders with characteristically cut-off abdomens beautifully patterned in red, brown and black are often found resting on the foliage of shrubs, or perhaps suspended motionless from a few disorganised threads. The handsome variegated colour pattern sets them apart from the two other kinds of spiders with this truncated form of abdomen, the Thomisid *Sidymella* and the full-coloured theridiid, *Episinus*. A closer examination of the front two pairs of legs will finally establish that these spiders are mimetids, with the unique arrangement of long erect bristles breaking up a row of shorter ones (Figure 109). These gay little spiders, which have been placed in Mimetus, are found all over New Zealand wherever there are low shrubs for them to inhabit. It has always been clear that they do not catch their prey with a snare. However, we have occasionally observed them on the webs of other spiders. The only record of them actually feeding was a single field observation made by C. L. Wilton who came across one eating another spider, one of the green crab spiders which, like the Mimetus, live in shrubs. These meagre observations could suggest that their habits are similar to those of their relatives in England where W.S. Bristowe has established that their favourite food is none other than their fellow spiders. We tried to obtain more definite information by keeping some specimens at home and then providing them with a variety of possible victims. The first thing we found was that they would not attack each other but then neither would they catch anything else. The only interest shown was in specimens of the theridiid *Achaearanea* which they approached with the long front legs extended but having touched the potential victim they did not carry the matter any further. What was interesting was the fact that *Achaearanea* did not attempt to catch and eat the mimetid even when it was creeping on its web.

Bristowe found that the English species, when provided with a *Theridion*, reached out rapidly to grasp it with the front pair of legs and then jerking its body nearer immediately bit its legs. Death came very quickly to the *Theridion* if it was bitten on the leg but not if the bite was on the abdomen.

Apart from the bright little species we have mentioned under *Mimetus*, there are many species living on the forest floor which are quite different from any other mimetids known although they still have the characteristic arrangement of long and short bristles on the front two pairs of legs (Figure 110). These are all dark brown in colour with the abdomen partly enclosed in a hard, shiny sheath and the abdomen,

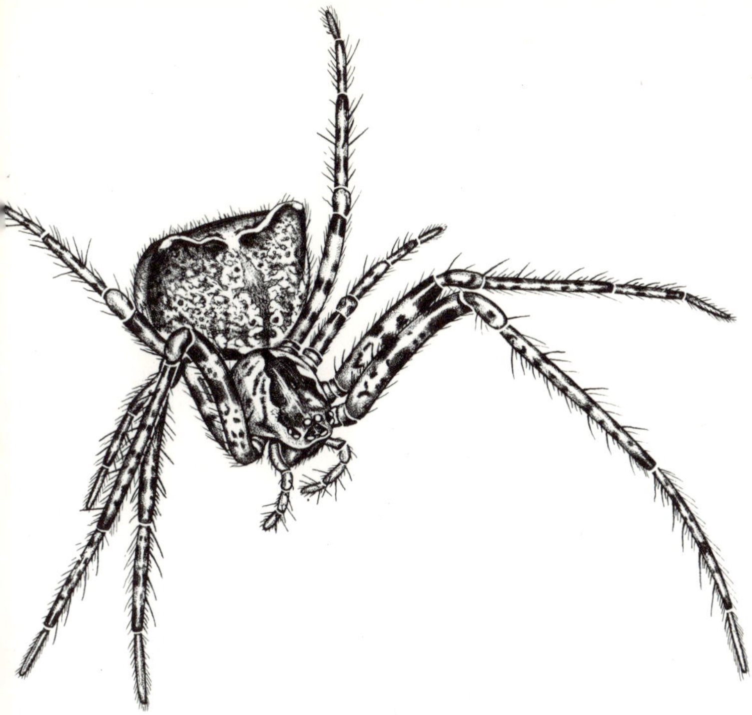

Fig. 109 *Mimetus* sp. (Mimetidae) female. The arrangement of the short and long bristles on the first and second pair of legs is similar in all mimetids.

Fig. 110 The forest floor mimetid.

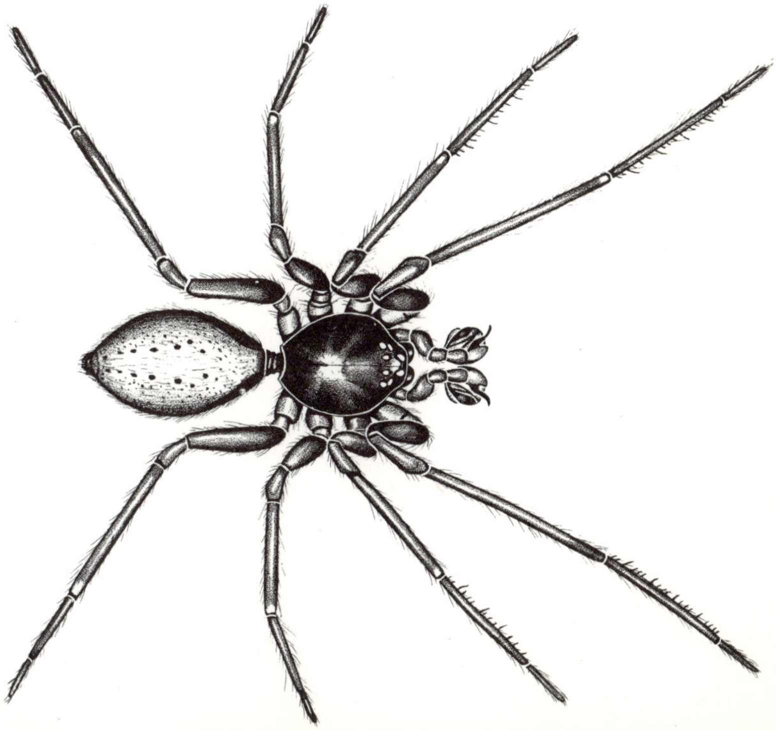

instead of being cut off, is a normal oval shape. These spiders move slowly over the forest floor at night and secrete themselves away in the dark during daylight.

It is thought that the mimetids may have sprung from the orbweb weavers; the mating habits follow the general pattern of this family. The male, approaching the female as she hangs from a thread, taps with his palps to provide the recognition signals. He then lays down a special mating thread on to which he coaxes his mate. According to Bristowe, the actual mating of the English species is very rapid with the male first moving along to the female, now suspended from the mating thread, inserting one palp and then retreating a short distance before approaching again to insert the other palp; the whole period of insemination lasts only a few seconds.

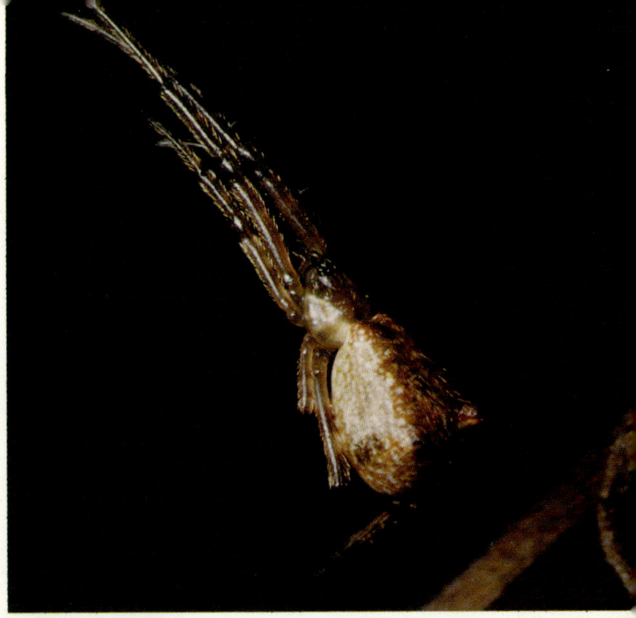

105. Top left. *Phoroncidia* sp. Female. Family Theridiidae. Waipori Gorge. Length 1.5 mm.

106. Top right. *Episinus* sp. Female. Family Theridiidae. Dunedin. Length 4 mm.

107. *Rhomphaea* sp. Male. Family Theridiidae. Nelson, Length 2 mm.

108. *Rhomphaea* sp. Female. Family Theridiidae. Stewart Island. Length 3 mm.

110. *Moneta* sp. Female. Family Theridiidae. Akatore, Otago. Length 2 mm.

109. *Rhomphaea* sp. Female. Family Theridiidae. Foxton. Length 3 mm.

111. *Argyrodes antipodiana*. Female. Family Theridiidae. Waipoua. Length 2 mm.

112. *Mynoglenes* sp. Female. Family Linyphiidae. Kaitangata. Length 4 mm.

CHAPTER TWELVE

Theridiidae: the Cobweb Spiders

This large group of small to medium sized spiders is often known as the comb-footed spiders because many of them use a prominent row of strong, curved and toothed bristles on the fourth pair of tarsi to manipulate the swathing silk with which they immobilize their prey. The characteristic web is an irregular network of strong threads of which only a few are sticky, and is commonly called a cobweb. Some however have developed specialized methods of catching prey which do not include the use of swathing silk and in these species the characteristic comb is reduced or even sometimes completely absent. The majority of the New Zealand theridiid spiders can be recognized and placed into a genus by their general appearance but there are some which require detailed study before they can be named and these we omit from the present narrative.

The best known theridiid is undoubtedly the notorious katipo which we have covered in the section on poisonous spiders, but some related, although harmless, relatives which keep appearing in homes and causing needless panic should be mentioned.

The spider most commonly mistaken for the katipo is the house cobweb spider *Steatoda grossa* which now enjoys worldwide distribution. The female is usually brownish with a characteristic pale half-moon patch on the front of the dorsal surface of the abdomen and pale areas behind while the male, which is more slender, usually has the pale patches broken up by transverse bands of brown. The untidy tangled webs which seem to collect dust in the corners near the ceiling may well become the bane of the housewife but the spiders themselves are rarely seen unless they become dislodged from the web or more commonly when the male is found striding rapidly about while looking for a mate.

A number of other species of *Steatoda* are found in more natural and undisturbed surroundings although some also make their homes in the garden. The abdomen of most of these spiders is shiny black, relieved by a few white patches and occasionally with a white crescent mark in front. It is not yet known whether they are truly natives of New Zealand or whether they are some of the known species from other countries. They are commonly found in grass and scrub near beaches as well as in dryer, open country inland but not in forests. Their webs are typical cobwebs constructed near the base of shrubs and tufts of grass or under loose-lying stones and debris.

Other, smaller black or brown spiders with pale patches on the abdomen belonging to the genus *Lithyphantes* also live in these habitats and even more commonly under stones in riverbeds.

Several species with a more or less globular abdomen belong to the closely related genera, *Theridion* and *Achaearanea*. These spiders are usually predominently brown, but some may be grey while many have a distinct scalloped band down the middle of the abdomen. *Achaearanea veruculata,* which is the species usually found in settled areas, often makes its web on the outside walls of houses (Figure 111). Throughout this book we have mentioned spiders which have been introduced to New Zealand from the northern hemisphere but *A. veruculata* is one species which has reversed this trend by becoming established in the

Fig. 111 *Achaearanea veruculata* (Theridiidae) female.

Fig. 112 Female Katipo *(Latrodectus katipo)*.

Scilly Isles off England, where it was probably introduced with plants sent from this country. *Theridion ampliatum*, with its distinctive pale brown carapace broken by a median band and uniform pale brown or cream abdomen, constructs its web close to the ground. The genus *Episinus* includes a large number of arboreal spiders which are easily recognised by the way the abdomen seems to be cut off behind where the surface slopes down steeply to the spinnerets (Figure 113). There are only two groups of spiders in New Zealand (*Sidymella* and *Mimetus*) with which they may be confused. The integument of the crab spiders of the genus *Sidymella* are however usually dull in appearance whereas in *Episinus* it is shiny. *Mimetus*, on the other hand, has a relatively shorter abdomen and is furthermore characterised by the striking arrangement of short and long bristles on the front two pairs of legs. The male of *Episinus* is much more slender than the female and the abdomen pointed at the end so that the cut-off section is not so clear but it can still be seen. Most of these spiders are yellowish or reddish-brown and sometimes the dorsal surface of the female is black. W.S. Bristowe described the simple web of an English species which constructed an

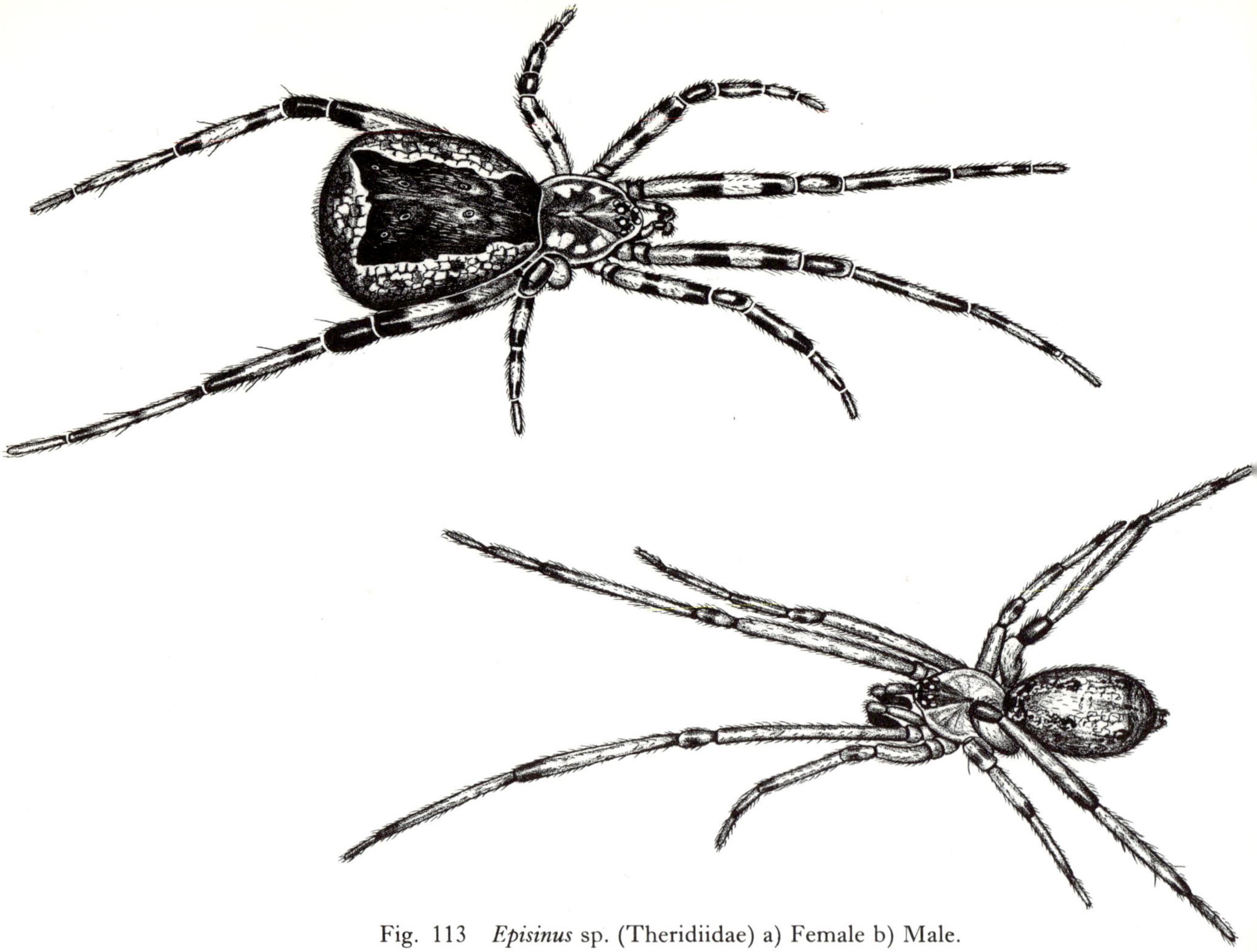

Fig. 113 *Episinus* sp. (Theridiidae) a) Female b) Male.

H-shaped frame from which the spider hung down with its hind legs clasped to the cross-thread, and the other three pairs holding on to the two lower arms. The two lower arms were sticky where they touched the ground and so the staple food of this spider was crawling insects such as ants, which blundered into the sticky threads. The New Zealand species of *Episinus* seems to be very similar to the British *E. angulatus* and it may be thought that their habits would be similar, but this does not appear to be so. The native species, which are found in forest or scrub, live well above the ground and construct a single line snare stretched between two twigs or branches. At night the spider is found hanging upside down from this thread with two pairs of legs in front and two behind but unfortunately we have never been able to observe any prey being caught.

While the use of a single thread snare by *Episinus* is still problematical, there is no doubt that this method is successfully used by at least some of our species of *Phoroncidia* (Figure 114). These spiders are all rather

Fig. 114 *Phoroncidia* sp. (Theridiidae) Female.

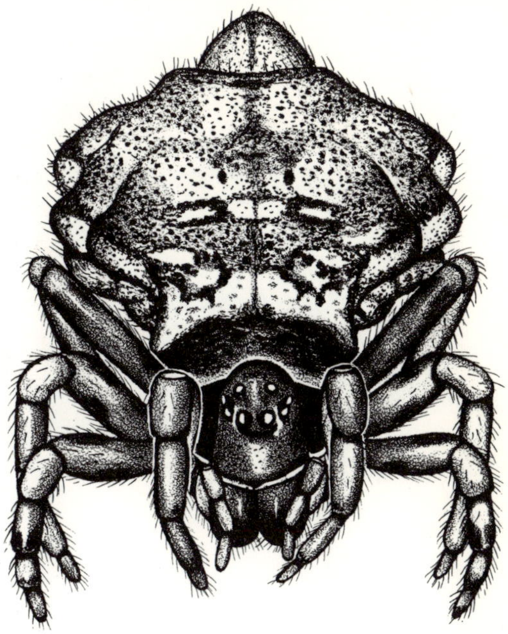

grotesque midgets, none exceeding two millimetres in length, and most smaller. The abdomen is always leathery and studded with minute shallow punctures while the skin is wrinkled and the surface drawn out into peculiar humps and mounds. To add to their unusual appearance, the head region of the male is often drawn up into a stalk, where the eyes are placed. The presence of these spiders has been known for a long time but their habits were studied only recently by Professor B.J. Marples. He found that they have reduced the snare to perhaps the ultimate in simplicity. A single thread bearing a few sticky globules is strung between two points by the spider which settles down at one end holding the thread taut with one of the front legs. As soon as an insect comes into contact with the globules, the spider releases the tension so that the prey becomes tangled in the slack thread. Then, pulling itself along the thread to the struggling insect, it lays a new snare as it goes.

Other minute theridiids, only slightly larger than *Phoroncidia*, also live in low shrubs and perhaps the most unusual of these are the various species of *Dipoena* (Figure 115). Actually there is nothing particularly striking about the females, but the male is extraordinarily different, because the carapace is monstrously enlarged, extending up into a relatively huge flat-topped box with the top surface creased and wrinkled. Overseas species have been reported to rely on a diet of ants and, although this might also be the case with our species, some are found in districts where very few ants occur and on that fact alone would need to vary their diet if they are to thrive.

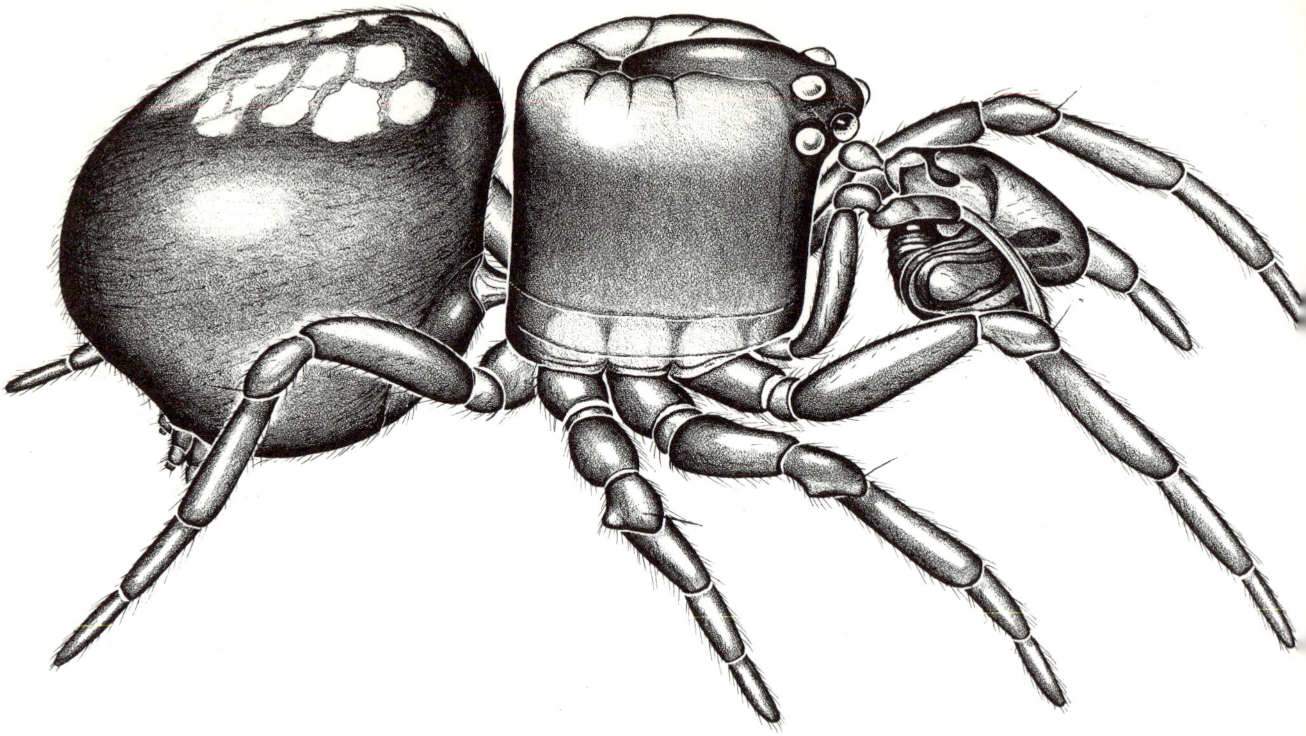

Fig. 115 *Dipoena* sp. (Theridiidae) Male.

A surprisingly large proportion of the New Zealand theridiids have the upper surface of the abdomen drawn up into a peak so that from the side the outline is triangular. The genus *Moneta* includes a number of species which do not entirely match this description because the dorsal projection is in the form of a hump but they can be looked upon as an example of the trend which has reached a peak in *Rhomphaea*. In *Rhomphaea* the abdomen is always higher than it is long and in some so extremely attenuated that it is difficult to imagine how the spider can move, until it is realized that it always moves upside down along a thread of silk (Figure 116). Occasionally these spiders are seen with a small tangle of threads but at night are usually found at the end of a single thread suspended horizontally in a shrub. Whether this is the snare, and, if so, how it catches prey, is not known.

Another numerous group of superficially similar spiders of the genus *Tekella* constructs a communal sheet web close to the ground in damp forest. These spiders are always darker than *Rhomphaea* and sometimes the abdomens may not be so strongly modified. The sheet web, which may be a foot or more wide, is built of extremely fine silk but it is in the tangle of threads above the sheet that the spiders normally live. Here twenty or thirty spiders, including young as well as adult males and females, are usually found suspended head downwards. No doubt there are many interesting facets to be explored regarding these spiders including the fact that no one is even sure that they are theridiids; at present they are mentioned here because we do not know where else to include them.

When the snares of orbweaving spiders are studied, the parasitic theridiid spider *Conopistha antipodiana* may also be seen. The first thing to catch the eye is the gleaming silver of the conical abdomen glistening like a drop of liquid quicksilver in the sun as the spider moves about the web of its huge host with seeming impunity. This striking little spider is also found in Australia and as it seems to prefer the webs of the Australian orbweb spider, *Aranea pustulosa*, it may have been an introduction to our fauna. Many species of these spiders are found in different parts of the world, particularly in the tropics, and all seem to live on the scraps left on the web by their compliant host.

Fig. 116 *Rhomphaea* sp. (Theridiidae) Female.

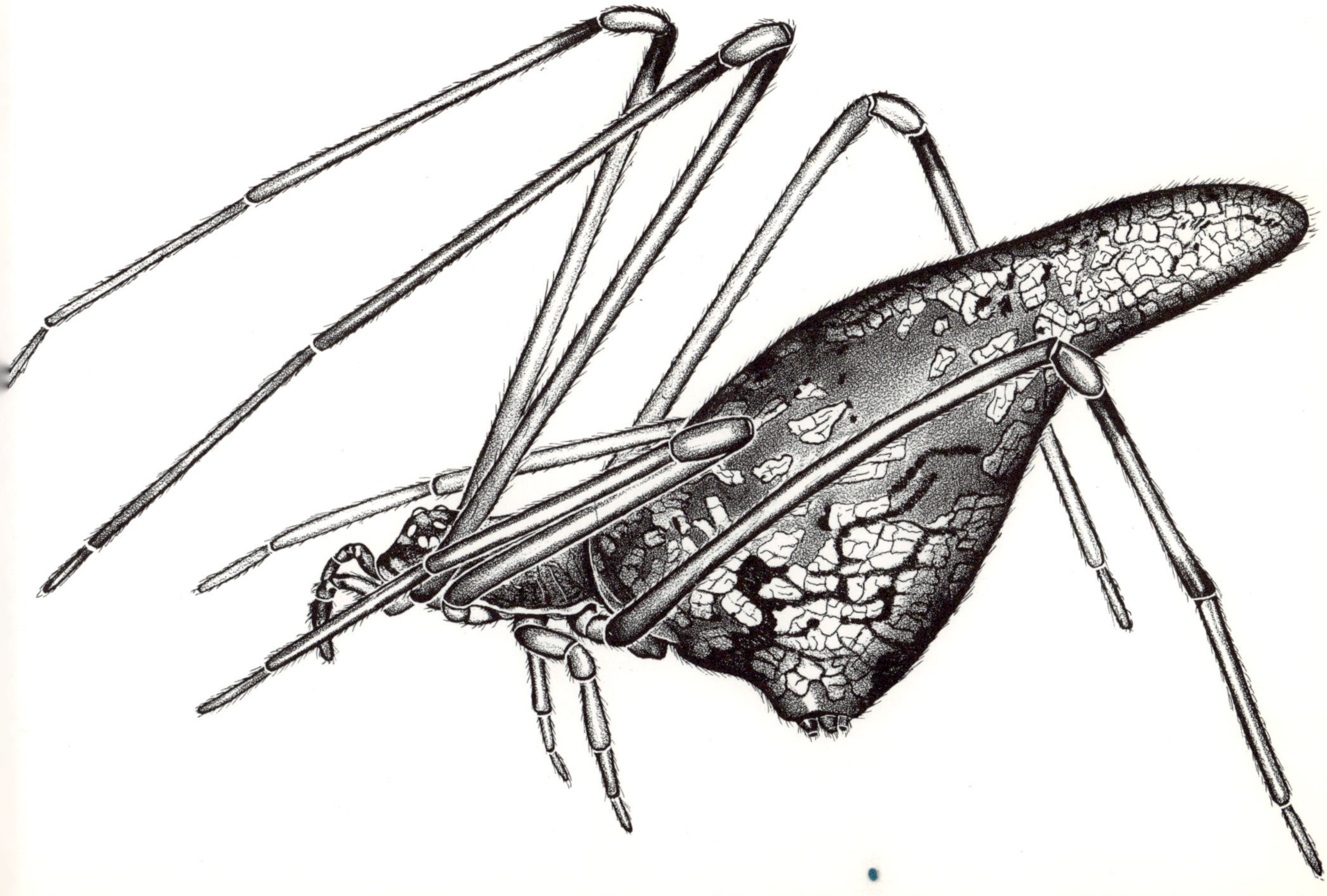

CHAPTER THIRTEEN

The Midget Spiders

Here and there, while discussing other families of spiders, we have mentioned some particularly small species such as the theridiid *Phoroncidia* or the pigmy male of the moth-catching *Celaenia* but there are three whole families of spiders in New Zealand in which the species are all minute and so smallness is part of their way of life. Two of these families – the Symphytognathidae and Archaeidae are represented only by native species which are mainly restricted to the forest but the third, the Erigonidae, all immigrants from the northern hemisphere, have in the absence of a strong open country fauna, taken over the grasslands and pasture and live in much the same way as in their original homes.

SYMPHYTOGNATHIDAE

Only twenty years ago, the symphytognathids were thought to be among the rarest of all spiders as only a handful of species or even specimens were known and these came from widely separated parts of the world. However, as interest has grown in the ecology of the forest floor cover – that world of moss, fallen leaves and debris – it has become evident that today the southern countries are their main homes. In New Zealand they are one of the most important elements on the forest floor where they play the role of predators of the smaller insects and other arthropods which abound there. At present it seems as if New Zealand is the main headquarters for the family and the probable reason for this is that, because of isolation, they have never been subjected to competition from more vigorous spiders such as the erigonids, some of which occupy similar habitats in the northern hemisphere. Whatever the reason, they deserve some mention in this book as being one of the most interesting sections of our forest fauna and as long as one can "think small enough" they are easy to find.

Smallness brings with it certain problems of which the most drastic is the loss of body moisture through the integument. As the animal becomes smaller, the area of the surface of the outside skin becomes proportionately greater than the bulk of the organs within, which means that a correspondingly greater proportion of water is lost through evaporation. Unless some positive steps are taken to reduce the loss of body fluids, it may dry out to such an extent that the internal body processes can no longer function and it will die. Many of the minute animals which live amongst the leaf litter in the forests avoid this fate by living only in places where the humidity is so high that there is little water loss through the skin. As a result, their continued well-being hangs on a slender thread as they are liable to perish if by some mischance their home dries out. A more positive method is shown in many of the spiders, including most of the symphytognathids, which have much of the abdomen covered with thick plates. In some the whole skin has become thickened and hard, so that water losses are reduced. Another way by which these spiders may lose moisture is through the open spiracles leading to the book lungs and tracheae but more particularly those whose book lungs, by their functional nature, present a large internal surface to the air. A tendency is shown in all true spiders for the replacement of the posterior pair of book lungs with tracheae but the symphytognathids carry the development further by a reduction in the number of leaves followed by their eventual elongation into tracheal tubes. So one of the most interesting characteristics of the symphytognathids is the complete loss of the book lungs in the great majority of species. These developments can be readily explained by logical reasoning but it must be admitted that experimental proof is so far completely lacking.

Another peculiarity of the symphytognathids is the tendency for the pedipalp of the female to be reduced in size and also number of segments so that in many species only the coxal segment is left as a seemingly useless knob, while in others even this has gone. That these spiders can manage without this pair of appendages is not particularly surprising because there are few of the normal functions of the pedipalp which cannot be taken over by the front pairs of legs; what does give food for thought is why other spiders have not also been modified in this way. The pedipalp of the male however, is normal in every way because, like all spiders, the male must use the complex palpal organ on the tarsal segment to transfer sperm to the female during mating.

The high carapace in both males and females of many species of symphytognathids sets them apart from other small spiders and most of the commoner New Zealand species belong to this group. The most widespread and numerically abundant species belong to the genus *Textricella*. These extremely small spiders, never more than a millimetre long, stretch their small sheetwebs amongst the moss and liverworts on the forest floor or on the trunks of forest trees. It is thought that they must feed mainly on springtails and other minute insects. Most of the species of *Textricella* are golden-brown but a few are black. The carapace is evenly raised so the outline from the side is almost a square while the anterior median eyes are always small and in some species have been lost so that only six eyes remain.

Chasmocephalon armatum is another very small spider, about the same size as *Textricella* but dark brown. A minute orbweb is constructed amongst the moss and litter. The abdomen of the male is protected by a hard shield on both dorsal and ventral surfaces with a further group of small plates at the side while the unprotected skin is thickened and spotted with small, hard spots. *Pseudanapis spinipes* is somewhat similar but more than twice as large while the head region is both higher and rounder (Figure 117). These spiders like *Chasmocephalon* probably use an orbweb to capture their food. They live on the forest floor but may also be found at the base of clumps of grass and sedge on swampy ground where the humidity is constantly high.

Fig. 117 *Pseudanapis spinipes* (Symphytognathidae) Male.

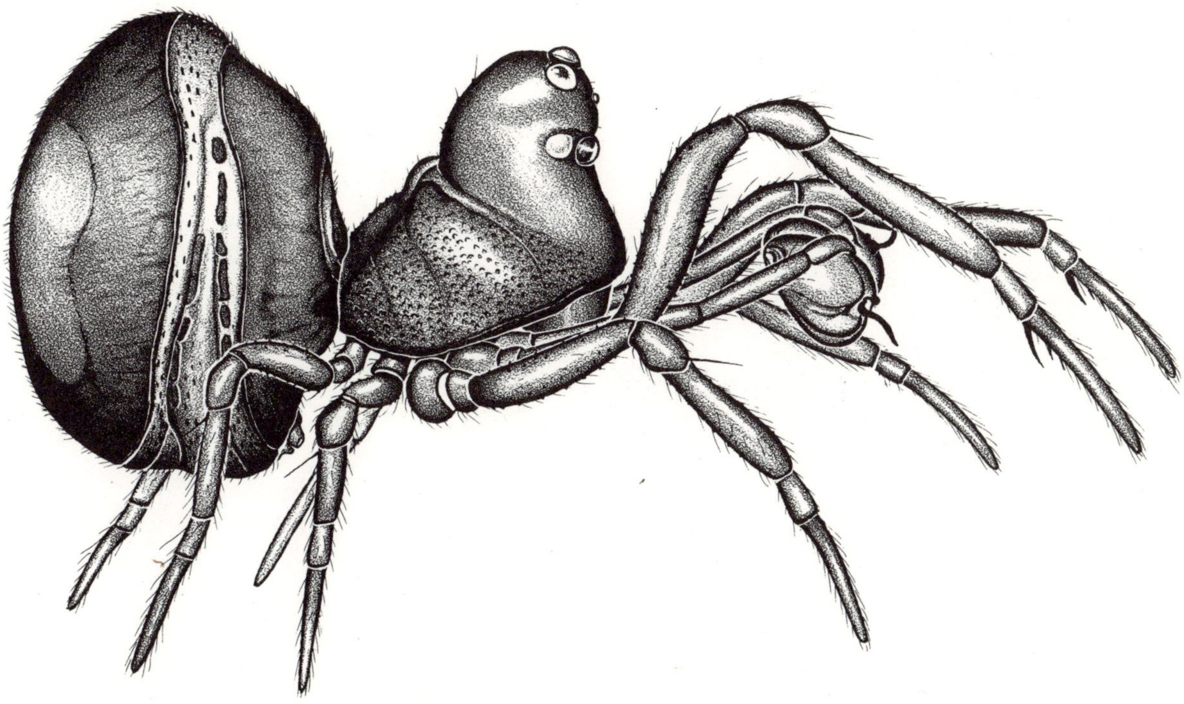

All of the species we have discussed so far are characterised by their high carapace and a tendency for the skin of the abdomen to become thickened but there are two other genera which, by contrast, are soft-bodied and delicate while the carapace is modified only in the male. Neither of these genera have been officially recorded as occurring in New Zealand although they have been known for some time and are worldwide in distribution.

Mysmena has rarely been found and then amongst the forest floor debris but it is less than one-half of a millimetre long so perhaps it is not surprising that it has not been seen more often. Some of the overseas species construct minute orbwebs while others apparently make a tangle web but nothing is yet known of the habits of the species which occur in New Zealand. The eye region of the males of *Mysmena* is slightly raised but in the related genus *Trogloneta* this tendency has been carried to an extreme with the eyes of the males placed on the end of a long and slender rod extending up from the carapace (Figure 118). It is difficult to imagine a use for this grotesque modification but when the spiders are studied it will probably be found that, as with the erigonids mentioned later, it performs some function during mating.

Fig. 118 *Trogloneta* sp. (Symphytognathidae) Male.

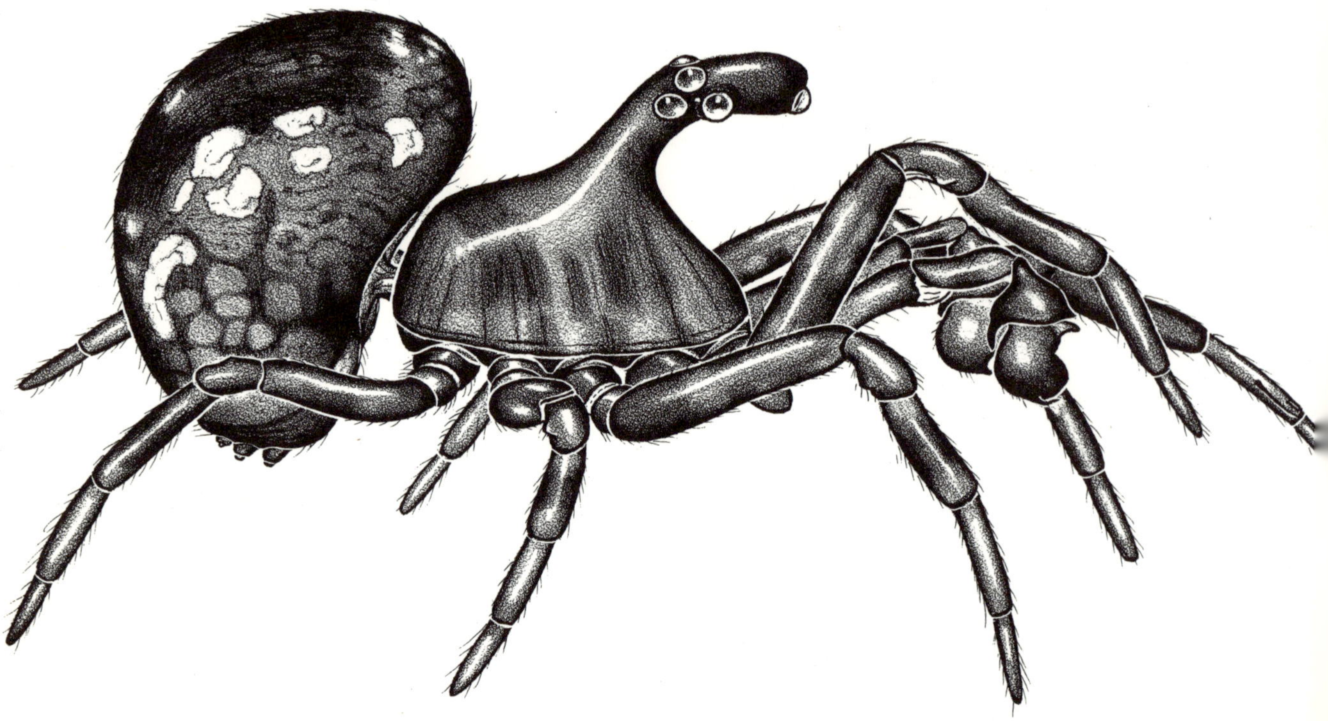

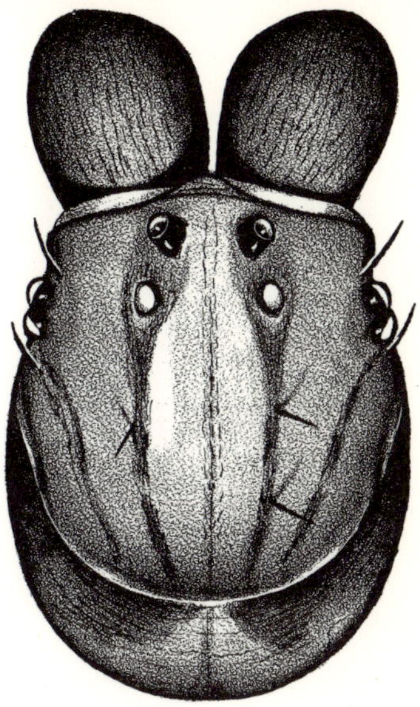

Fig. 119 *Pararchaea magna* (Archaeidae) Carapace.

Trogloneta is found in a variety of habitats all with a humid atmosphere such as low ferns and shrubs in the wetter rain forests, moss and debris on the forest floor, the carpet of mosses and bryophytes clothing the trunks of trees and also in clumps of sedges and grasses growing in marshy ground. It has been noted that the web is little more than a tangle of threads but these spiders have never been closely studied.

ARCHAEIDAE

The family Archaeidae has an interesting background because it was first established in 1854 for six fossil species which had been found imbedded in pieces of Baltic amber. These specimens were still in such a good state of preservation that one could think that the period which had passed from the time they were trapped in the gum oozing from the trunks of these trees growing in the Baltic Region of Europe until they were found could be numbered in scores of years instead of the thirty millions or more years which had truly passed. These spiders were all small, with the carapace grotesquely elevated, particularly in the head region but, unlike those of the symphytognathids, the chelicerae had moved up with the carapace and remained near the eyes and as the fangs of all spiders must be able to reach the mouth the chelicerae were correspondingly long (Figure 120). The family was thought to represent yet another evolutionary experiment in specialization which had long since died out leaving no living descendents, and then in 1881 the Reverend O. Pickard-Cambridge described a living species from Madagascar. Over the next eighty years a score or more species of archaeids were found in the southern Hemisphere, from South America, South Africa, Madagascar, Australia and finally New Zealand, but it is certain that the once extensively distributed representatives in the northern hemisphere have now all perished. The only studies of the

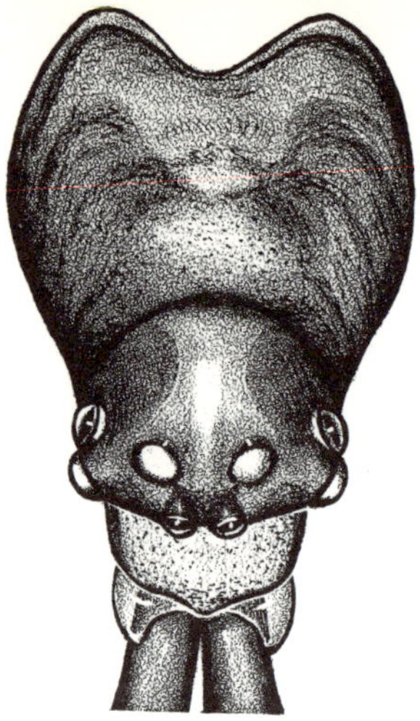

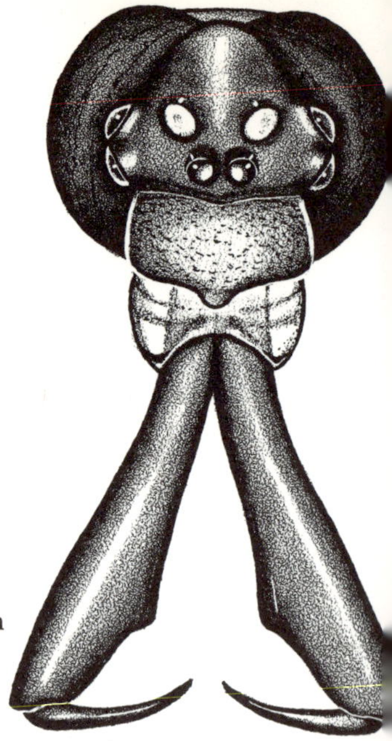

Fig. 120 *Holarchaea novaeseelandiae* (Archaeidae) a) Carapace from above. b) Chelicerae and head from in front.

habits of these spiders have been concerned with the Madagascan species, and these, like the mimetids, are said to prey on other spiders. Of the three genera found in New Zealand, *Holarchaea* with a body-length of less than a millimetre is by far the smallest (Figure 121). *Holarchaea novaeseelandiae*, the only known species, has been found in forests down the west coast of the South Island from near Nelson to Fiordland where it lives on low growing ferns and in moss. Three species of *Zearchaea* have been recorded of which *Zearchaea clypeata* from the southern part of the North Island and the east coast of the South Island is the most widely spread. This delicate little spider, about one-and-a-half millimetres long, lives amongst the low foliage near forest margins and in fern but at least in the Wairarapa seems to prefer the accumulations of dead leaves in the bases of niggerheads which grow in swamps and marshes. A distinct overhanging ledge in front of the carapace distinguishes this genus from the similar *Pararchaea*. *Pararchaea alba*, with its distinctive creamy white abdomen, has been found in many parts of the South Island, but always among the moss on the forest floor, while *P. rubra* with its equally striking bright orange-yellow abdomen is characteristic of the North Island forests.

As none of these spiders has ever been found associated with a snare, it is possible that they are all hunters; whether they prey on other spiders like their distant relatives in Madagascar has still to be discovered.

ERIGONIDAE

The spiders we mention here are the little shiny black spiders which as children we called money spiders because it was thought that to do one of these active little fellows a good turn would eventually result in a stroke of fortune mostly to be thought of in terms of hard cash. They were also sometimes called rain spiders on the understanding that if one was accidentally squashed then rain would follow. It is interesting to read the accounts by W.S. Bristowe of the long held superstititons in rural England because it is clear that these tales were brought out by the early English immigrants along with the spiders, all of which have been introduced.

The most abundant species is *Diplocephalus cristatus* which has taken over pasture-land throughout New Zealand, and it is this minute, shiny black spider which is often seen running over the lawns and paths about our homes (Figure 122). The small sheetwebs these spiders construct to catch their equally small prey of springtails and other insects are found amongst the grass close to the ground. The head of the male is raised and divided into two distinct lobes but this species can be considered most conservative when compared with some related species with extraordinary spines, mounds and bumps. While *Diplocephalus* has not been observed mating it is known that in other species this

Fig. 121 *Holarchaea novaeseelandiae* (Archaeidae) Female.

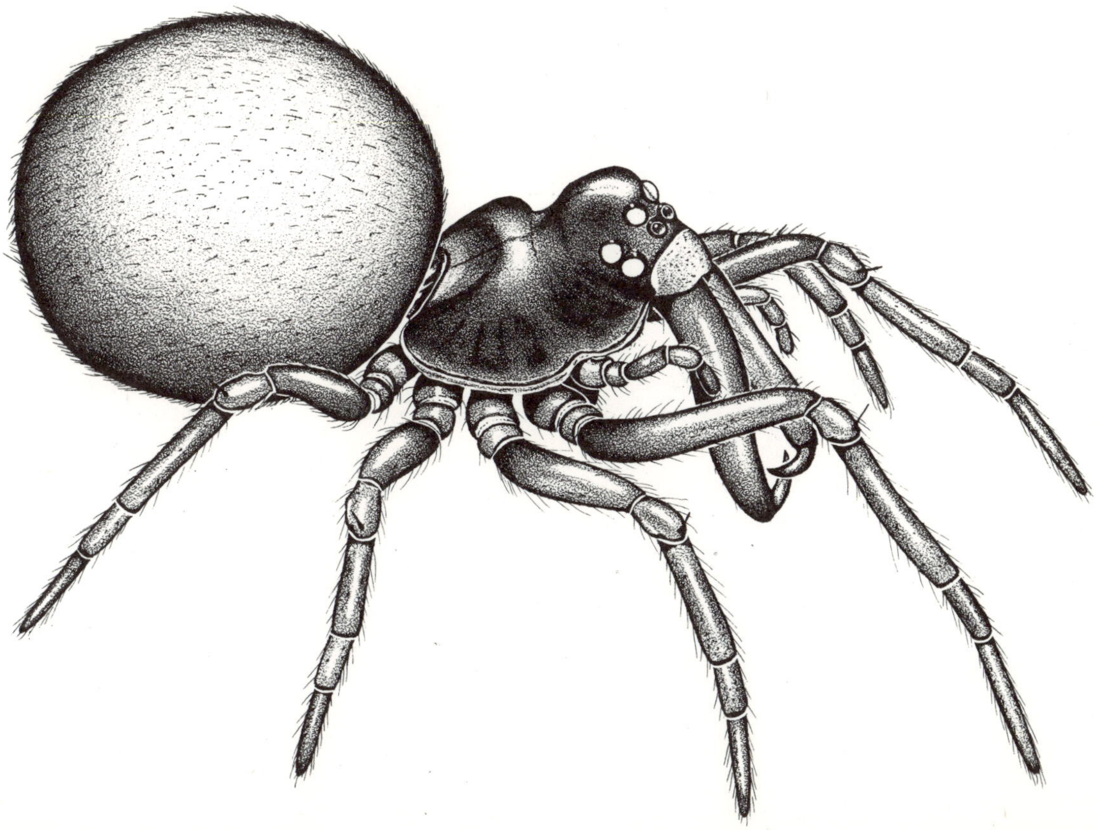

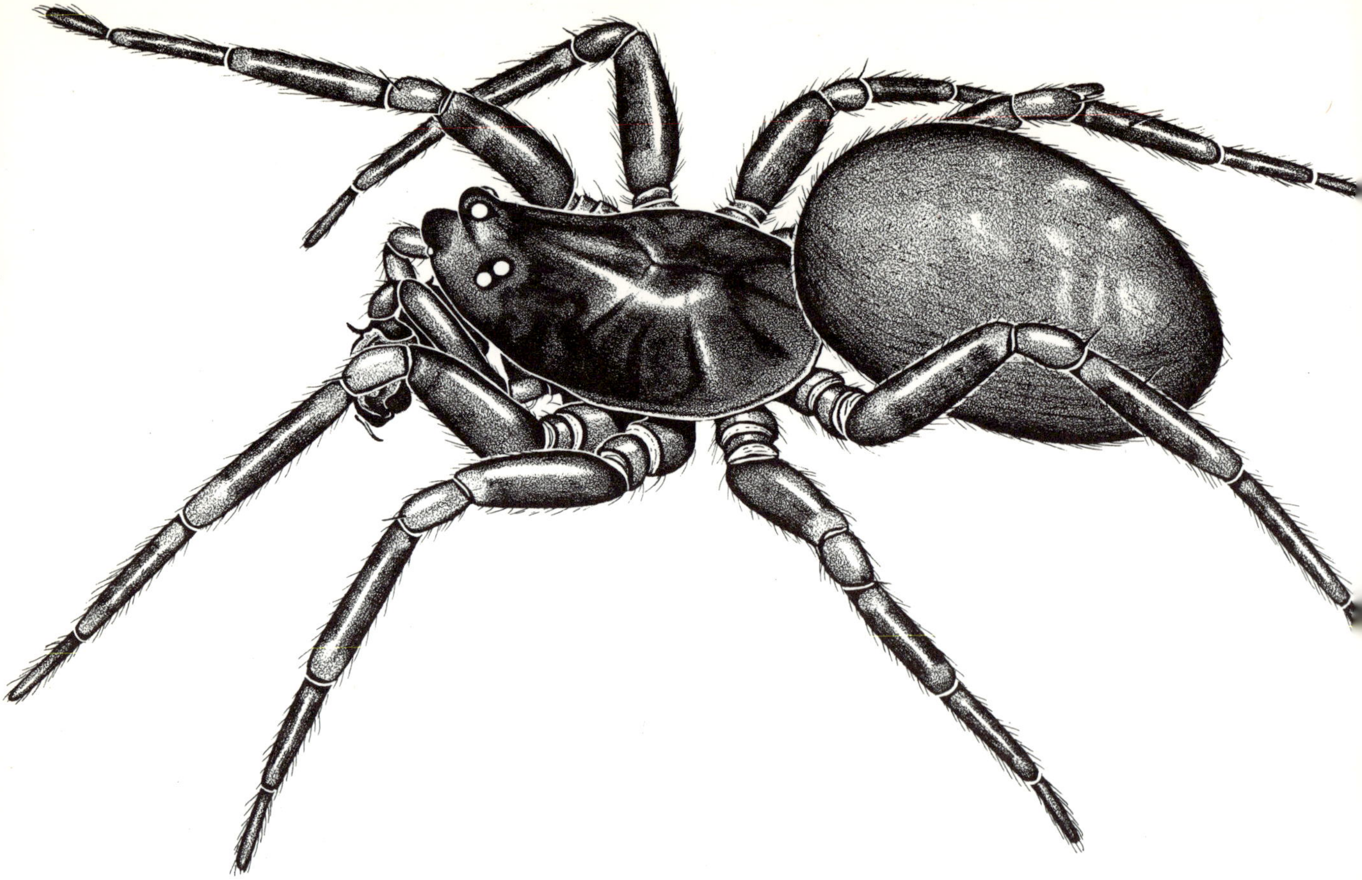

Fig. 122 *Diplocephalus cristatus* (Erigonidae) Male.

structure is held by the chelicerae of the female as mating is accomplished. The second known group of these spiders in New Zealand belong to the genus *Erigone*, in which the head is not modified but instead the outer margin of the carapace of the male is studded with small teeth to give it a saw-toothed edge. These small spiders are also black and are found in the pastures with *Diplocephalus*; they, too, are common on the sandy beaches where *Diplocephalus* is never found.

LINYPHIIDAE

The spiders of this widespread family are known generally as sheetweb builders and although the webs of our species are not very noticeable, they usually take this form. The majority of species are plain grey, black and brown and have never aroused much interest. Some specialists consider that the erigonid spiders should also be placed in this family but, as the New Zealand species are easily separated, we have described them under the separate family name. The largest linyphiids in this country belong to the two genera *Mynoglenes* and *Paralinyphia*, which

113. *Synotaxus* sp. Female. Family Linyphiidae. Dunedin. Length 2 mm.

114. *Synotaxus* sp. Male. Family Linyphiidae. Gore. Length 3 mm.

115. *Tekella* sp. Female. Family Theridiidae. Fox Glacier. Length 2 mm.

116. *Cambridgea foliata.* Female. Family Agelenidae. Feilding. Length 25 mm.

117. *Cambridgea* sp. Female with egg-sacs. Family Agelenidae. Dunedin. Length 15 mm.

construct small sheetwebs near the ground beneath lying wood and stones. The species, which are all difficult to tell apart, are coloured with blacks and browns, although some may have a greenish tinge but as with many other linyphiids the abdomen, being only sparsely clothed with hairs, is shiny. During the summer, a single egg-sac is constructed near the web in the form of a low mound covered with a thin layer of gleaming white silk through which the eggs can be seen (Figure 124). While these spiders are common in the forest, they are also able to stand exposure in grasslands and are even abundant in the windswept sub-antarctic islands. Many of the linyphiids which inhabit the bush have long slender legs and a small narrow body and look very much like the pholcids except that the eyes are placed in two rows of four instead of being bunched into two triads and a single pair. *Synotaxus crinifrons* is typical of this abundant but undescribed fauna. In the male, the surface between the eyes is raised into a prominent mound provided with a bunch of drooping hairs. The web is a beautiful inverted dome surmounted by an extensive tangle of threads amongst which the spider hangs.

Fig. 123 *Mynoglenes* sp. (Linyphiidae) Female.

Fig. 124 *Mynoglenes* guarding her egg-sac.

At least two species of linyphiids have successfully colonised this country. *Ostearius melanopygius* is a conspicuous, orange coloured spider with a contrasting black patch at the end of the abdomen. In New Zealand it is found amongst rough grass in open country particularly in the north. The spider has a most peculiar history. It was first described in 1879 by the Reverend O. Pickard–Cambridge, the outstanding English spider authority of the last century on specimens sent to him from New Zealand; so it duly took its place in the lists as a native New Zealand spider. Some twenty years later it was found living in England and it was then naturally assumed that here was a case of a New Zealand spider having reached England and become established. Eventually, it became clear that the opposite explanation was true and that the species represented yet another English introduction to New Zealand which, after becoming established thousands of miles from its homeland, had been sent back to be described by a specialist living only a few miles from where it originally came. The second English species *Lepthyphantes tenuis,* recently identified by Mr C.L. Wilton, has a distinctive black abdomen with white patches and is widespread in pastures throughout the country.

CHAPTER FOURTEEN

Agelenidae

Many New Zealand species are placed in this family because they share a number of simple characters. They are all three-clawed, non-cribellate, spiders with a well developed colulus at the base of the spinnerets. The eyes are usually in two rows of four, the male palp has a prominent projection at the side of the tibia, the legs are strongly armed with spines and there is a blister-like swelling at the base of the chelicerae where they meet the margin of the carapace. At present it is easy to examine a specimen and say that it is an agelenid but, unfortunately, detailed studies of the internal anatomy show that our spiders represent several different groups which have become superficially similar, and so in future some of the spiders mentioned below will be placed in separate families, but they may still be recognised from the notes and illustrations.

One spider which will undoubtedly remain in this family is the ubiquitous house spider *Tegenaria domestica*, which while originally native to the northern hemisphere has become worldwide in its distribution. As its scientific name suggests, it has become closely associated with man and is rarely found far from buildings (Figure 126). The web is a thick horizontal sheet leading to a short tunnel retreat wedged into a crevice or a corner. In middle or late summer the dingy white egg-sacs are attached to the surface within or close to the retreat. The body is slightly less than half an inch in length but the legs are relatively long and hairy. The carapace is light brown but usually darker than the abdomen, which has five indistinct chevrons on the posterior half of the upper surface. When an insect alights on the sheet, the spider dashes out from its retreat to capture it and then drags it back to the retreat to eat it. It is worth nothing that this spider walks over the upper surface of its sheetweb because this is in direct contrast to the other spiders with which it might be confused, e.g. the native *Cambridgea*, which always

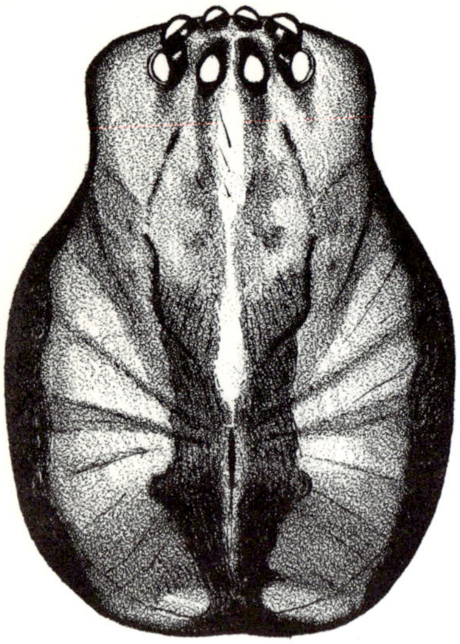

Fig. 125 Carapace of *Cambridgea antipodiana.*

moves upside down beneath its sheet. Although it is probably found all over New Zealand, *Tegenaria domestica* seems to thrive better in the north and becomes more rare further south. We thought that it might be useful to include some of the more pertinent aspects of its life history but we found that as is often the case with those animals which have become an accepted part of our lives no one has apparently studied it in detail, so the necessary facts were not available. One further characteristic feature of *Tegenaria* is that the hind pair of spinnerets are conspicuously longer than the other two pairs and with one exception this serves to distinguish the house spider from all other agelenids in

Fig. 126 The introduced house spider *Tegenaria domestica.* Female.

Fig. 127 Forest floor agelenid related to *Tegenaria*. Male.

New Zealand. The only exception is a group of small spiders whose body length is little more than a quarter-of-an-inch and which build small sheet webs under loose bark and logs in the forest (Figure 127). These active little spiders, mottled with black and sometimes a reddish flush on the top of the abdomen, lead a secluded life and so perhaps it is not surprising that they have not been officially described and their habits have not been studied.

The commonest agelenids comprise a number of species of *Cambridgea* some of which have been able to extend out from their normal forest homes and become firmly established in our gardens and even under houses and in disused sheds. The webs are conspicuous with a finely woven horizontal sheet drawn out below into a number of tent-like inverted domes by strong anchoring threads and topped by an extensive maze of irregularly placed lines (Figure 131). During the day the web is deserted but at night the spider creeps out from the darkest recesses of the sheet to cling upside down on it. All species of *Cambridgea* look very much the same with conspicuous black bands down the middle and sides of the carapace and some degree of black mottling on the abdomen but they differ greatly in size. The very small species with a body length of a quarter-of-an-inch o; less are still restricted to the forests and it is the larger species which grow to three or four times this size which are commonly seen in gardens. The snare is without sticky threads brt they are obviously not needed because after an insect has flown into the knockdown threads and dropped on to the sheet it has little chance of escape. The spider immediately detects the movement of the web and, moving rapidly upside down beneath the sheet, reaches the victim and thrusts through the silk to bite. Sometimes if an insect becomes tangled in the upper threads the spider will move around the sheet to capture it as it struggles and then bring it down to the lower surface. The poison of these spiders is extremely effective as even large insects succumb within a few seconds. No swathing silk seems to be used and the prey is taken to the back of the web to be eaten. The webs of the largest species, *C. foliata* (Figure 129), from the North Island may be up to three feet wide but those most commonly seen are perhaps one third

Fig. 128 *Cambridgea antipodiana*. Female.

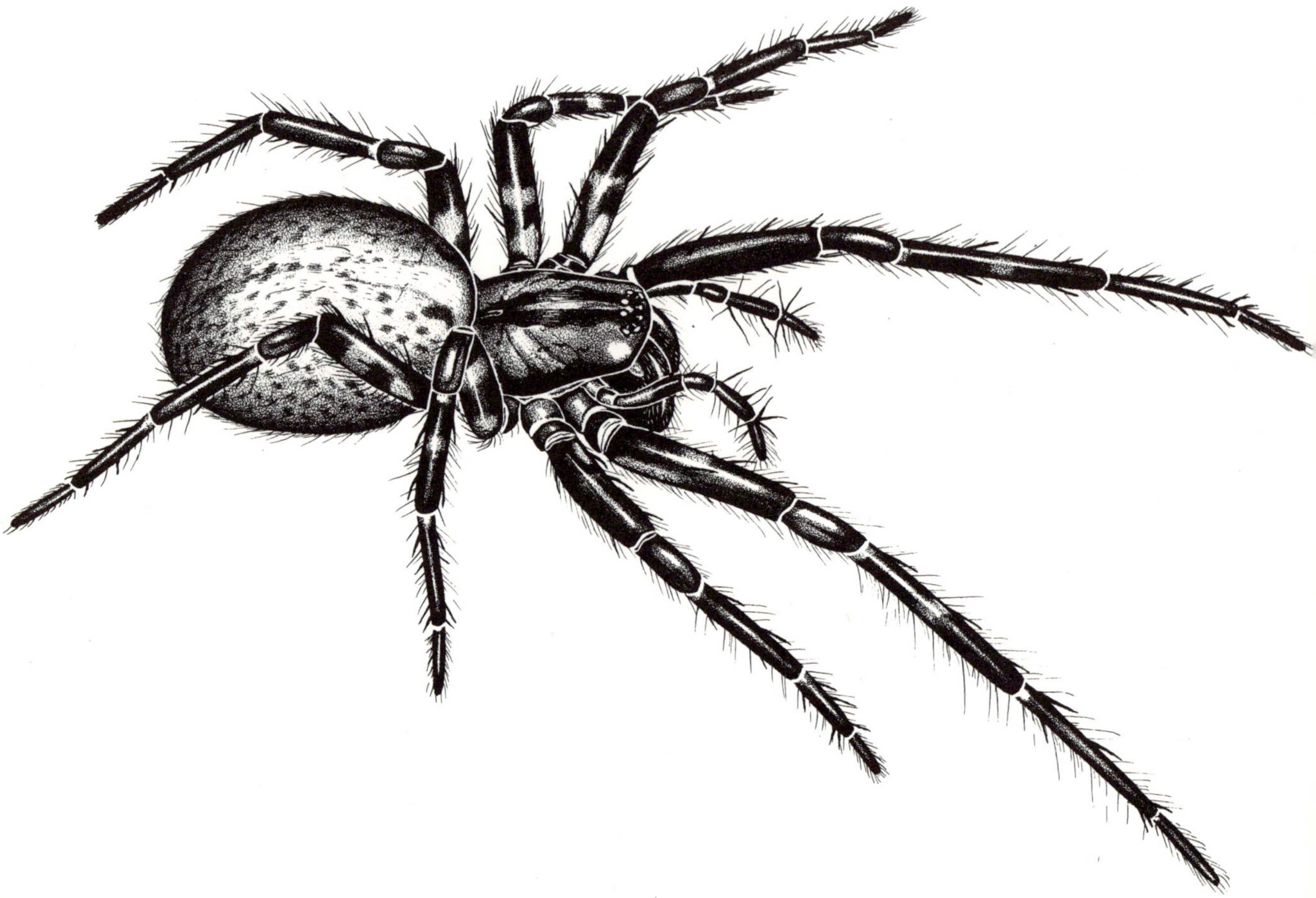

Fig. 129 *Cambridgea foliata*. Female.

of this size. There is another difference in that *foliata* always constructs its web from a proper tubular retreat which it makes in a hole in the trunk of a tree. Our other species do not use such a definite retreat so it is interesting to note that the related Australian *Corasoides australis* which constructs an identical web near the ground, does so from a tunnel in the ground which may be up to eight inches deep. The egg-sacs of *Cambridgea* which may number three or four are hung in the web near the retreat but are not always easy to see because the conspicuous white silken cover which actually encloses the eggs is hidden under a thick coating of debris which the spider has hauled up in its fangs from the ground (Figure 132). The mating of these spiders has never been observed but a curious feature of the males of some species is a strong projection from the waist which seems to rub against a hard ridged plate on the abdomen. This is suspected to be a stridulating organ which the spider uses to produce special vibrations when it approaches its mate (Figure 4). The chelicerae of the male are usually very much larger than those of the female; so it is possible that they are used at this time to

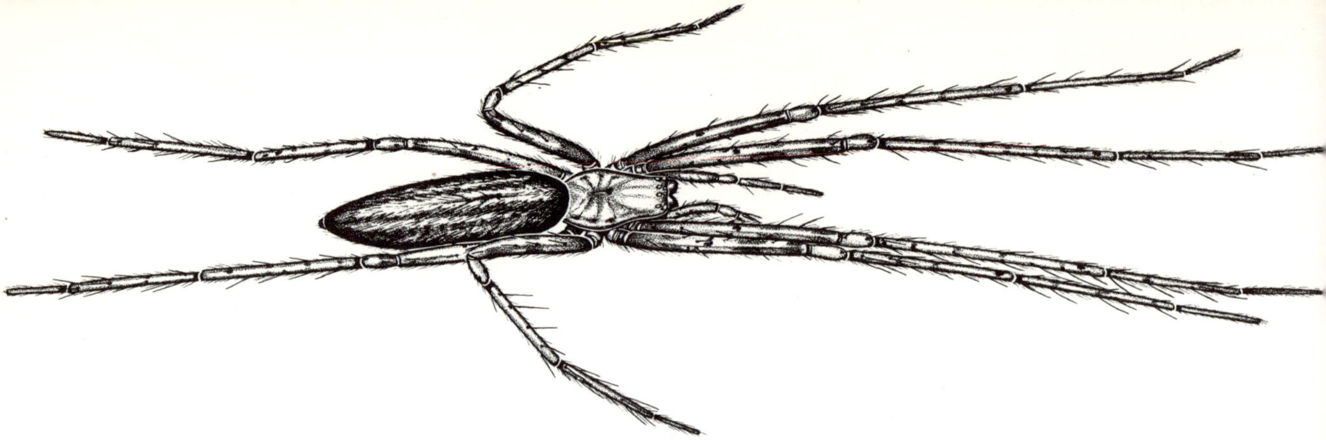

Fig. 130 *Ischalea spinipes*. Female.

grasp the female, perhaps in much the same way as by *Tetragnatha*. The extraordinarily flattened and elongate *Ischalea spinipes* has long been known but no one has been able to decide where to place it (Figure 130). At times it was thought to be related to the elongate orbweb spiders, the Tetragnathiids, and at others regarded as an unusual looking nursery-web spider but in fact it is a close relative of *Cambridgea*, having abandoned its snare to become a hunter on the foliage of trees and shrubs The usual colouring is pale straw brown but sometimes pale green specimens are found. These spiders rest with the front two pairs of legs stretched out in front, the back pair out behind, and the third pair bent to the side and seem to prefer lying along the midrib of a leaf so that they are difficult to see. At night when they hunt, their movements are slow and deliberate and it is difficult to imagine them catching an active insect but, as we have failed to keep them alive in captivity we do not know how they go about it.

THE SEA SHORE SPIDERS

One of the most fascinating of our spiders is the marine species *Desis marina* which is found only in the narrow stretch of shore around our coasts between the high and low tide levels. This handsome spider some eight millimetres long with its bright reddish brown carapace and contrasting cream abdomen is striking enough but what really stands out are its huge chelicerae which are almost as long as the carapace (Figure 134). The first intimation that a truly marine spider occurred in this country came in 1877 when Mr C.H. Robson, the resident Cape Campbell lighthouse keeper at that time reported on his observations to the Wellington Philosophical Society. Mr Robson, like so many of the people who take up these lonely jobs, was an ardent and observant naturalist and we can do little better than repeat the notes he recorded at this time.

'I found a veritable spider at home under the water having a nest in an

Fig. 131 Sheetweb of *Cambridgea*. Normally only one sheet is present.

Fig. 132 Having constructed the egg-sac *Cambridgea* adds a layer of debris.

old *Lithodomus* hole, of which the rocks about here are full. All of the spiders of this kind which we have found have had nests in these holes, and always under water at all times of the tide. Over the mouth of the hole the spider spins a close web, which when finished looks like a thin film of isinglass, and is waterproof; and behind this film is the nest and egg-sac, which last is of various shapes and contains a large number of eggs!

'When the spider is disturbed it goes to the bottom of the pool and if a small stick or straw is extended to it, it at once gets ready for a fight, advancing its long and powerful fangs for that purpose . . . When a small fish is placed in a bottle of water with one of these spiders, the latter will attack at once, driving its long, sharp falces into the fish near the head and killing it instantly!'

When the paper containing this information was published, Dr James Hector added a note in which he compared the habits of the marine spider with the well known Water Spider of England *Argyroneta aquatica*, and after concluding that these two spiders were related, proposed that the name *Argyroneta marina* be used for the New Zealand species. The next year Dr Llewellyn Powell, the doctor in Christchurch who had become interested in spiders through his studies of the poisonous katipo, pointed out that the spider was not related to the English spider but to spiders with habits similar to the New Zealand one from New Guinea and Singapore described under the generic name *Desis*. He suggested the marine spider be known as *Desis robsoni* so that, as well as being correctly placed, the name of the discoverer would be associated with the spider for all time. Unfortunately, in spite of the praiseworthy intentions of Dr Powell, the rules governing the allocation of scientific names insist on the acceptance of the first specific name bestowed and so the spider became known as *Desis marina*. The matter did not rest here because in 1879 the Reverend O. Pickard–Cambridge, the outstanding English authority at that time, deciding that our spider deserved to be placed in a separate genus, proposed a new name *Robsonia marina* and so for a brief period Mr Robson was again honoured. Alas, no other spider authority would agree that the New Zealand spider was so very different from the other known marine spiders and, as they would not agree to accept this new generic name, the spider became known once again as *Desis marina*; this is the ruling which stands today. Little has been added to our knowledge of these spiders since the original observations made ninety years ago other than that they live on most rocky shores around the coast of New Zealand and the offshore islands and that casual observations indicate that their staple food consists

Fig. 133 Male *Cambridgea* capturing a drone fly.

Fig. 134 The inter-tidal spider *Desis marina*. Female.

of small crustaceans such as the amphipods which abound where the spiders are found.

There is another group of agelenid spiders associated with the seashore which were first brought to notice by P. Goyen, an early school inspector in Otago. Goyen published an account in 1889 of a spider he had found living under stones below high water in Otago Harbour. He called this spider *Habronestes marinus* and had the following comments to make:–

> This interesting spider is found on the under surface of stones between the high and low watermark, being at full tide immersed in from a few inches to a foot or more of water. When detached from the stones it rises to the surface like a cork and at once simulates death. In this condition it will bear a good deal of pushing about without showing the least sign of life; but a slight pinch with the forceps makes it either run smartly along the surface of the water or strenuously struggle to dive. It, however, never succeeds in getting more than its head and some of its legs under water . . . To get under water it must have some support for its legs . . . I have also found examples of this spider in the crevice of the banks laved by the water at full tide, but, though I have searched diligently for it, have never met it any great distance from the sea.
>
> Its cocoon which is of a plano-convex shape, and attached firmly by its plane surface to the underside of stones, is also under water during a large portion of every day and they young, as soon as they leave the cocoon, seem as much at home in that element as the mother spider. The material of the cocoon is of a thin leathery consistence and apparently impervious to both air and water. It contains about twenty rather large pale – yellow, roundish eggs, which are placed in tiers, and built into a rude sort of a cone. They are loosely agglutinated together and attached to the central part of the roof of the cocoon, which is invariably well inflated with air. The young remain in the cocoon for some time after they are hatched, and this imprisoned air is no doubt intended for their use. Owing to its whitish colour the cocoon is a rather conspicuous object; but its fabricator is in colour so much like the bits of wood and other material adhering to the stones that it is difficult to distinguish it from them. It constructs neither snare nor protective tube that I could discover. The air by which its existence is supported during immersion is held entangled in the dense pubescence of the abdominal ventral surface.

Most of the observations made by Goyen have since been confirmed except for the curious fact that these spiders have never again been found living below the high tide level; so it may have been that his

Fig. 135 The seashore spider *Myro marinus*.

observations were made at a time of an exceptionally high tide. We have found the spiders in Otago Harbour as well as the coast nearby, together with their conspicuous cocoons but always amongst the debris and stones above the high tide level where they feed mainly on the sandhoppers. Other similar, but as yet undescribed, species have now been found living along much of the coast of both North and South Islands. It has been known for some time that the Otago Harbour seashore spider does not belong in the genus *Habronestes* in which Goyen placed it, but only recently has it become clear that it should be called *Myro marinus* and so related to the *Myro kerguelenensis* found only on some of the windswept subantarctic islands of the southern seas.

Somewhere near one hundred different species of small to medium sized greyish, black or brown agelenid spiders may be found on or near the forest floor in different parts of this country but, as very few of them have been described, we can do little more than mention that they are there. The two groups which have been recorded are the largest and most

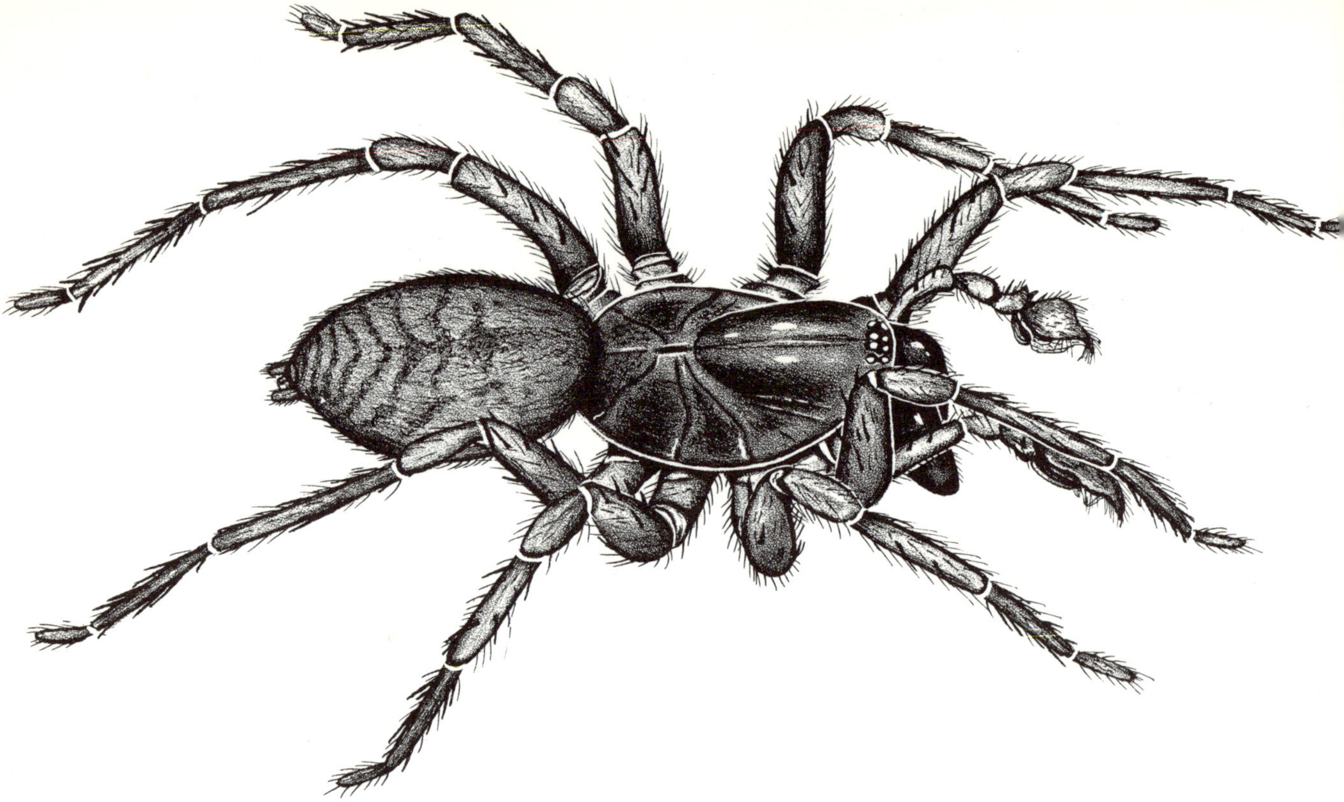

Fig. 136 The male of one of the common forest spiders at present called *Rubrius*.

Fig. 137 *Hahnia* sp. Female. All of these spiders have the six spinnerets placed in a row.

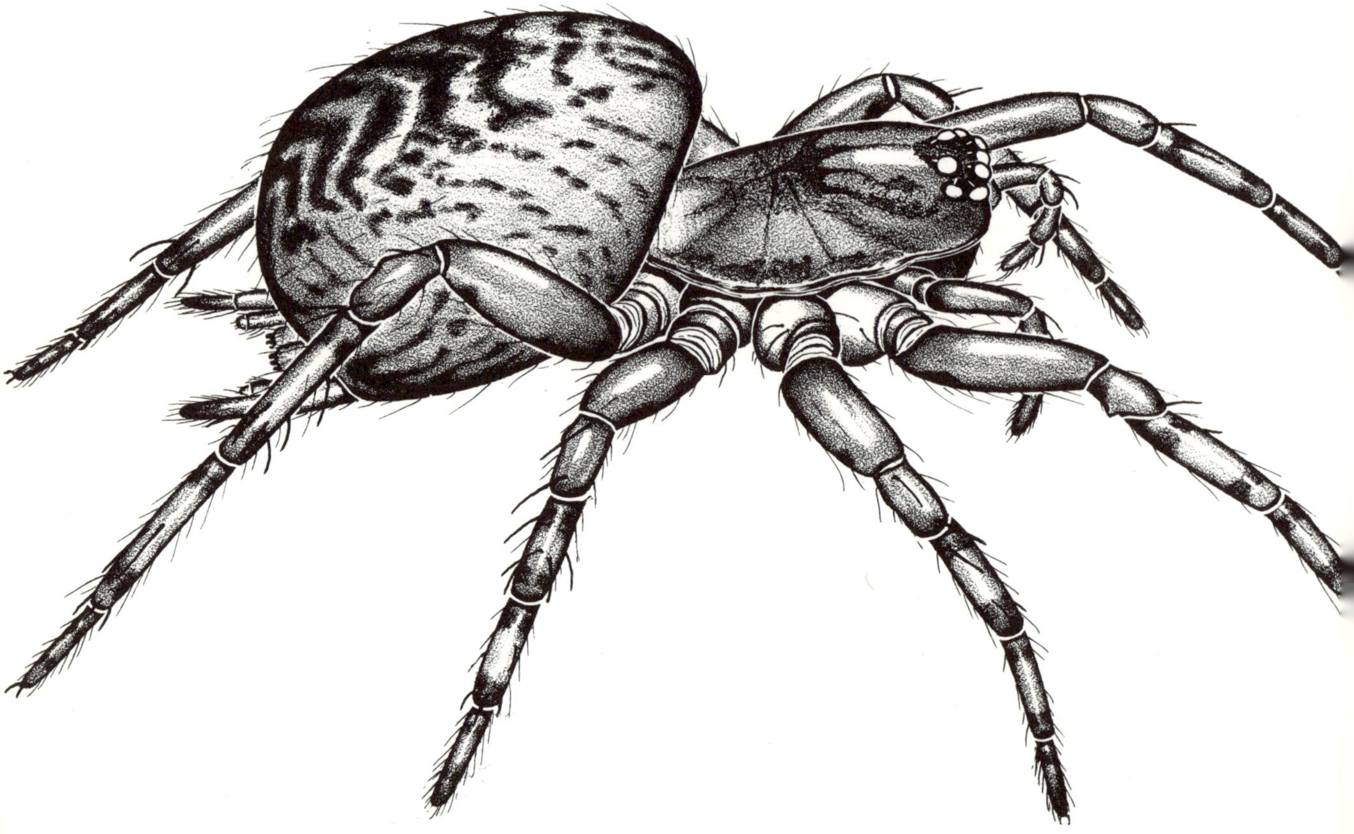

conspicuous of them all. The celebrated French authority on spiders, the late Eugene Simon, described one of these as *Amphinecta decemmaculata* as far back as 1897. This is a blackish brown spider, more than half-an-inch long, with ten small, pale patches arranged in five pairs down the dorsal surface of the abdomen. It lives in and under rotting logs on the forest floor with a very small sheet of silk but at night it may be seen wandering about and so it may well catch its prey hunting. During summer, the females are found secreted in a silk-lined retreat guarding a single, large and round egg-sac clothed in a loose cover of pure white silk.

The second group of similar sized spiders is usually placed in the South American genus *Rubrius* (Figure 136). These spiders are also dark brown but instead of pale patches the abdomen has a row of darker chevrons. Sometimes the young spiders are found sheltering in delicate silken tubes open at each end but it is probable that these retreats are used only during moulting and that, like *Amphinecta*, these spiders actively hunt for their prey. The female guards her egg-sac until the spiderlings hatch but, unlike that of the *Amphinecta*, this is shaped like a purse and is strung from the surface beside her.

There is a group of midgets usually less than two millimetres long associated with the agelenids but often placed in a family of their own – the Hahniidae. The main character which separates these little spiders from other agelenids is the remarkable arrangement of the spinnerets where the whole six are placed in a single transverse row across the abdomen (Figure 137). Only one of these spiders has been recorded

Fig. 138 *Amaurobioides maritima*. Female.

from New Zealand. *Hahnia solitaria* from Canterbury, but there are a score or more from all parts of the country waiting to be named. Most of these are greyish, rather squat in appearance with a few dark chevron stripes on the abdomen but some are pale yellow or even red. All live on or near the ground where they construct a small sheetweb beneath which they lurk. Few observations have been made on native species but the mating position taken up during fertilisation by the male of some of the European species is most unusual as he stands to the side of the female but facing the same way while he reaches beneath her abdomen. It would be interesting if someone were to study some of the New Zealand species which are quite different from their northern relatives just to see if these also use the same method.

AMAUROBIOIDES

It is unusual to have a family of spiders devoted to a single species but this seems to be the case with the family Amaurobioididae. The single species, *Amaurobioides maritima*, with its conspicuous dark brown pattern, was first described by the Reverend O. Pickard-Cambridge in 1883 from specimens sent to him from Allday Bay in Otago. Subsequently further species were described from South Africa, Campbell Is. and Tasmania but these are now thought to be the same as the New Zealand species.

In New Zealand, *Amaurobioides* may occur on any part of the coast where conditions are suitable. The spider lives in a tube of tough silk constructed in a crack on the rock face or perhaps on a small ledge where sand has accumulated but always near high tide level within reach of the seaspray, the region called the splash zone. Just why these spiders should be restricted to this habitat is puzzling because it has been found that they still thrive when in captivity well away from the sea and then are not particularly selective regarding their food. In their normal habitat, however, they seem to feed mainly on the Isopods which abound in cracks and crevices and so perhaps it is their preference for this prey, which is undoubtedly restricted, which in turn limits the spread inland of the spiders.

The tube is usually open and it is the white inner lining which catches the eye when one is searching for the nests but during the period after Christmas when the eggs are laid, the opening may be covered in with silk. The egg-sac is lens-shaped and attached to the inner wall.

118. *Cambridgea* sp. Immature male. Family Agelenidae. Kaituna Valley. Length 15 mm.

119. *Cambridgea* sp. moulting. Family Agelenidae. Punakaiki. Length 20 mm.

120. Arboreal agelenid. Male. Family Agelenidae. Fox Glacier. Length 8 mm.

121. ***Rubrius*** sp. Female. Family Agelenidae. Catlins. Length 12 mm.

122. ***Rubrius*** sp. egg-sac. Family Agelenidae. Waipori Gorge.

123. *Myro marinus*. Female. Family Agelenidae. Matanaka Beach. Length 6 mm.

124. *Dictyna* sp. (related to *D. rufomaculata*). Female. Family Dictynidae. Feilding. Length 2 mm.

CHAPTER FIFTEEN

The Cribellate Spiders

A number of spiders found in New Zealand possess the structures necessary to produce the particular kind of opaque thread which is known as cribellate, hackled or calamistrated silk. It is thought that ancestral spiders had four pairs of spinnerets and that in some cases the middle front pair became modified to form the cribellum, a flattened plate perforated with small holes from which exude a viscous fluid used to coat each thread produced by the normal spinnerets (Figure 139). All spiders which have a functional cribellum also have one or two rows of evenly spaced hairs on the metatarsus of the fourth pair of legs, known as calamistra (singular calamistrum) used by the spiders to comb out the hackled threads as they are produced. Although the spider families sharing these easily seen features are often grouped together under the convenient title 'cribellate' spiders they are not necessarily closely related to each other. Five families of these spiders are found in New Zealand but the bulk of the species fall into two families – the Dictynidae and Amaurobiidae.

DICTYNIDAE

In the northern hemisphere these spiders are abundant and, once you know what to look for, their webs are easy to find but the New Zealand species are both small and more retiring. All the native dictynids are placed at present in the huge worldwide genus *Dictyna* but, although there are two very different groups to be seen, both share the peculiar bow-legged structure of the chelicerae in the male form (Figure 140). We suspect that these unusual chelicerae have the same function as with their overseas relatives where the male uses them to clasp the chelicerae of the female so that they are held firmly while he inserts his palpi. The first *Dictyna* recorded from New Zealand was collected by a Frenchman

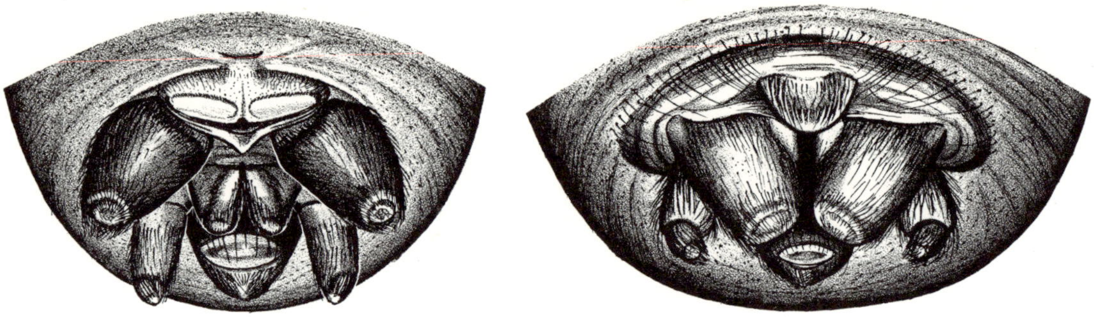

Fig. 139 The tip of the abdomen from below showing a) Spinnerets and cribellum of *Ixeuticus martius* b) Spinnerets and colulus of *Cambridgea antipodiana.*

Comte de Dalmas who collected spiders for his friend Eugene Simon of Paris while on a fishing holiday in this country before the First World War. Actually Comte de Dalmas became so interested in the spiders of New Zealand that on his return to France he decided to study the collection himself and subsequently published his large paper Araignées de Nouvelle Zélande which for the last fifty years has been the standard work on our native species. In this work he described *Dictyna cornigera* which he collected from the banks of the Temuka River in Canterbury (Figure 141). This is a small, greyish spider about four millimetres long with white hairs on the carapace and abdomen closely matching the colour of the river-bed stones. We now know that this spider is common on river-beds throughout New Zealand but is difficult to find until you become familiar with its web. The retreat

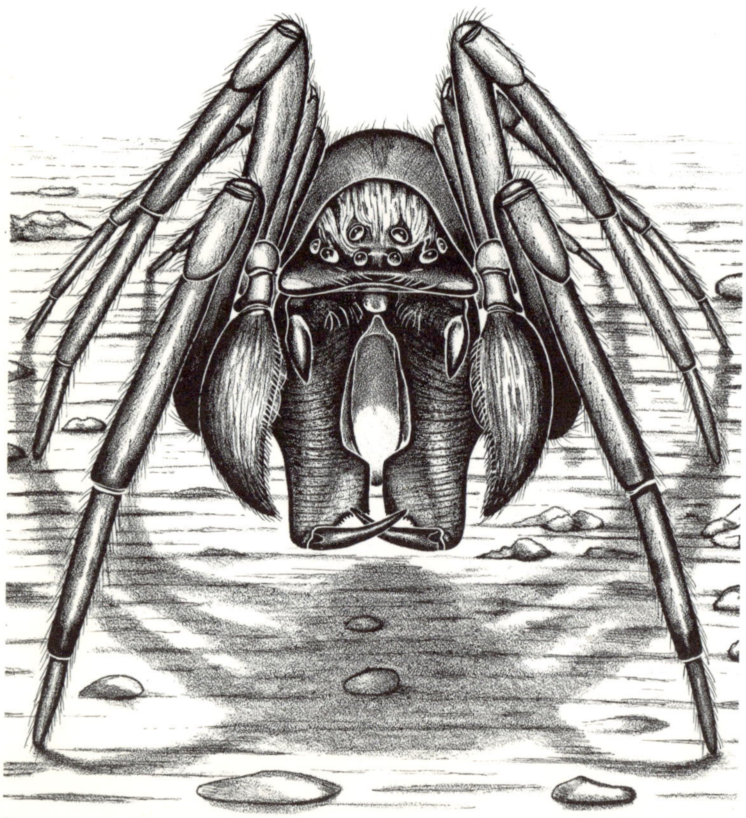

Fig. 140 The male of *Dictyna cornigera* has bowed chelicerae with which it grasps the female during mating.

is usually in the crevice between two closely spaced stones from which a number of radiating lines extend, bridged irregularly with a zig-zag thread of typical cribellate silk. The egg-sacs are round, loosely bound with white silk and guarded by the mother at the end of her retreat. The second group of species is very different in appearance as well as in habits. These spiders are various shades of green often with some white streaks and patches on the dorsal surface of the abdomen and some are further embellished with a conspicuous patch of red or brown (Figures 142, 143). Of the dozen or so species found in different parts of the country only one, *Dictyna rufoflava*, from Waiheke Island, characterised by having the red patch at the end of the abdomen, has been described.

All these spiders live openly on the surface of the leaves of low shrubs where they capture their prey by means of a sheetweb across the concavity of the leaf. The eggs are contained within a flat disc of silk plastered to the leaf beneath the web and held firmly in place by a number of short radiating bands (Figure 143).

Fig. 141 Female of the riverbed dictynid *Dictyna cornigera*.

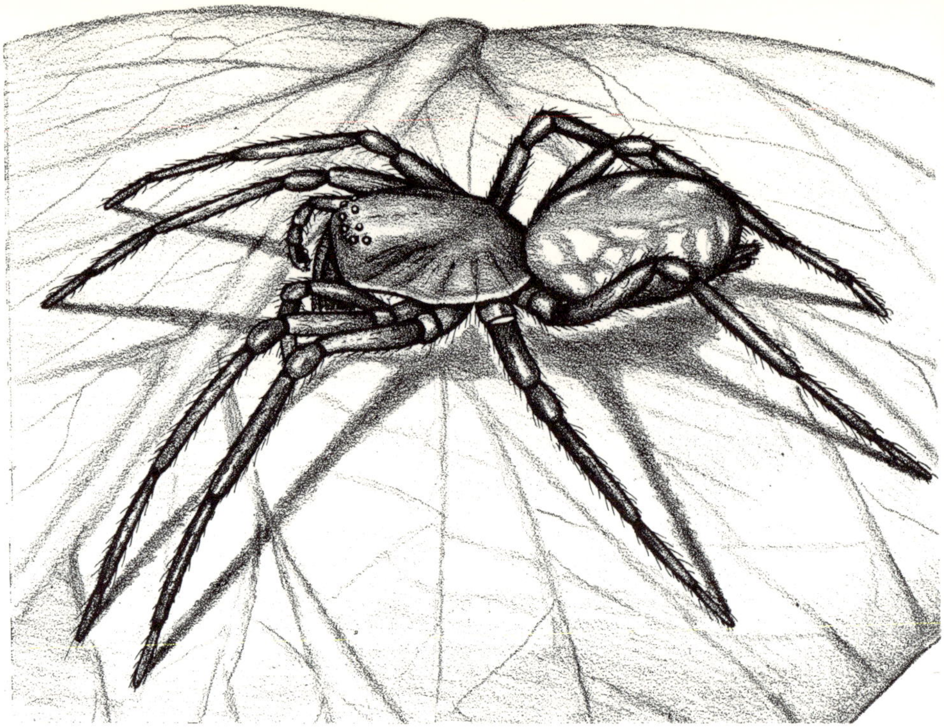

Fig. 142 The Green *Dictyna* lives on shrubs.

Fig. 143 The Green dictynid with its egg-sac.

Fig. 144 The Black House Spider *Ixeuticus martius*.

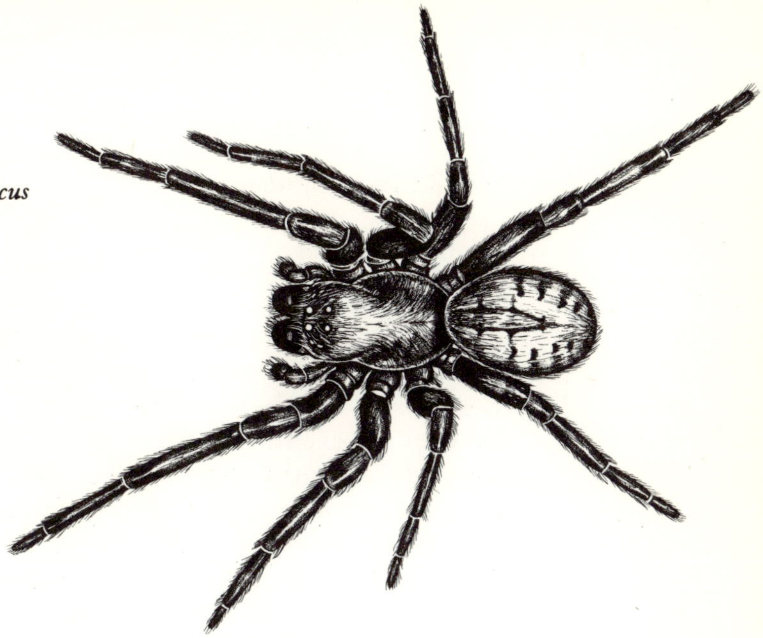

AMAUROBIIDAE

Apart from the presence of the cribellum and calamistrum the amaurobiids, mostly grey or brown spiders, are similar to many of the cribellate spiders at present placed in the Agelenidae, and together with this latter family constitute the bulk of the spider fauna of our forests.
The most commonly seen species, however, is the Australian *Ixeuticus martius* whose webs are conspicuous on the walls of houses, in hedges and shrubs and on rock faces in the more dry and open parts of the countryside. When first spun the web extending from both above and below a short retreat shows clearly the zigzag ladder construction so common in this family but as the web becomes older and the tenant keeps adding more threads it resembles a rather dusty sheetweb (Figure 145).

Fig. 145 The web of the Black House Spider.

During the day the spider remains in its retreat, although it will often come out if an insect gets caught or even when falling rain vibrates the threads. At night it takes up a position in readiness at the mouth of the retreat with its front legs stretched out on the web, dashing out quickly to capture insects trapped from time to time. The remains of these meals are deposited at the end of the retreat and so it is easy to discover by examining these empty corpses and pieces that the insects most often caught are beetles, flies and moths. These spiders have been studied little in New Zealand in spite of their abundance and home-loving habits but no doubt the following excerpts from the account by Mr N.L. Roberts of the mating behaviour of the closely related *Ixeuticus robustus* will largely apply to the species which has taken up residence here.

> A female spider, considerably larger than the male, advanced a little and began twanging the threads of the web. The male who was about three inches away, gave similar tactile responses, although not so frequently or vigorously as the female. After a brief interchange of these vibratory 'messages' the male suddenly walked round the back of the web and up into the main funnel shaped opening where the female moved to face him. Twanging was again resumed and the male evidently not certain of his reception, retreated twice when the female advanced slightly. Gradually confidence returned, and he advanced until they were able to touch each other with the first pair of legs. The legs moved backwards and forwards more rapidly, as if the spiders were 'sparring' until a stage was reached at which time the female suddenly became limp, the legs retracting and the body remaining inert as if supported only by the web. The male immediately advanced and applied his palps rapidly, after which he retreated through the funnel to the front of the web. The female soon became active again . . .

These actions were repeated a number of times and each time the female became comatose before the actual mating was accomplished. The genus *Oramia* includes a number of species which were at one time placed in *Ixeuticus*. These spiders are much less hairy than *Ixeuticus martius* and usually larger. All have a shiny reddish brown carapace with a tendency for the eye region to be heavily shaded with black while the abdomen tends to be grey-brown with pale areas on the dorsal surface, sometimes taking the form of a cross (Figure 146). While many species are found in the forest a number make their homes under loose stones in the open country and a few are found above the bush-line on the mountains of both the North and South Islands. Some in the southern portion of the

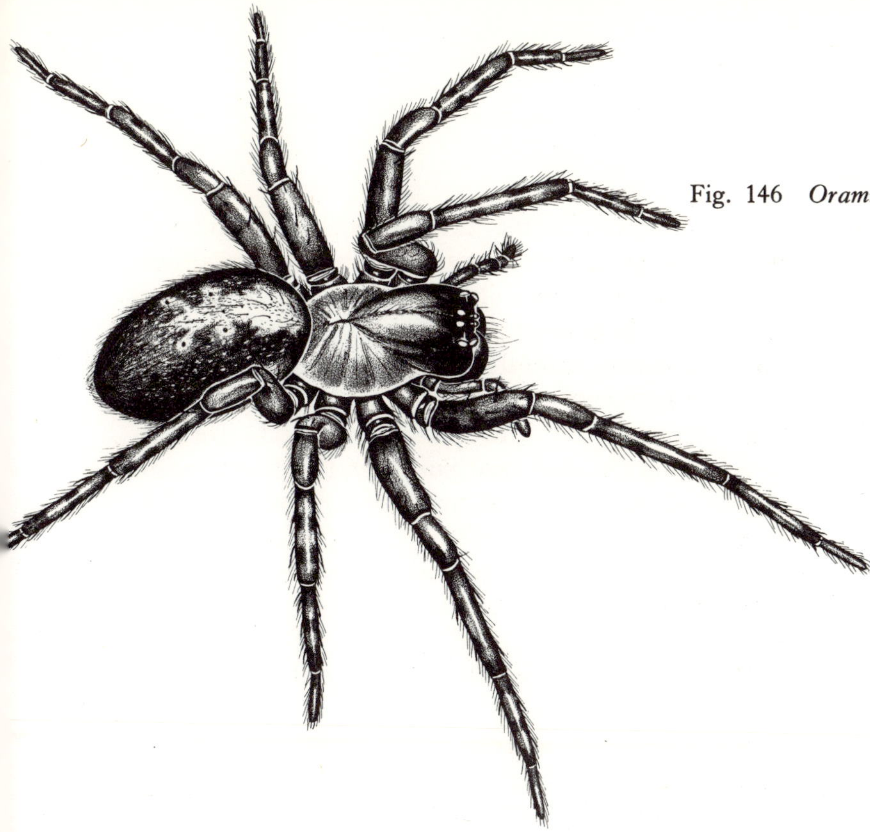

Fig. 146 *Oramia* sp. Female. From Fiordland.

Fig. 147 *Oramia* sp. with egg-sac. Dunedin.

South Island and the subantarctic islands have become specialists, restricted to a narrow zone amongst the rocks piled up on the seashore above the high tide level. As some thrive in open country it is not surprising that these spiders have become familiar denizens of gardens in the South Island. Those species, usually characterised by a pale abdominal cross, construct a small web of radiating lines bridged with the usual zigzag cribellate thread. The spider remains in its retreat beneath the web and here in middle and late summer the female can be found guarding a single, domed egg-sac attached firmly to the substrate by numerous fine threads (Figure 147).

A number of other native species are still retained in *Ixeuticus*, where they will remain until new genera are established; as the webs they construct are most unusual some mention should be made of them here. *Ixeuticus dalmasi* lives on the trunks of forest trees where it constructs a conspicuous white sheet web. These webs, about two inches across, are guyed on to irregularities of the bark so that the sheet is raised from the surface and it is here beneath the web the spider rests waiting for its prey to alight (Figure 148, 149). The egg-sacs are found spun on to the bark and because the young do not seem to wander far before setting up house for themselves the trunks of some of the trees may be quilted with many scores of webs, both large and small under each of which a spider lurks. *Ixeuticus dalmasi* is found in the southern half of the North Island but other species which construct similar webs are found in most parts of New Zealand. R.R. Marples described the web of another spider which he called *Ixeuticus janus* found on the overhanging faces of the limestone cliffs at Hanging Rock in Canterbury. Fundamentally the web was the same as that of *dalmasi* but a further sheet constructed beneath the surface sheet curved in to the centre to form a tube leading to the rock surface so that from the side the whole web was mushroom shaped. He found that this spider did not rest on the rock surface but between the two sheets of silk and it was here the disc-shaped egg-sacs were suspended.

We mention this spider because, by removing the outer sheet, it is easy to progress to the type of web made by *Stiphidion facetum*, a very different spider with conspicuously long legs and a mottled, greyish body which is also abundant in Australia (Figure 150). The web of this spider is also found on the under-surfaces of over-hanging banks and cliffs and consists of a single sheet, supported be numerous guy threads to keep it off the surface. However the sheet slopes in at the centre to form a wide, open funnel which extends to the rock surface, where the spider rests with its legs outstretched. The name 'Sombrero spider'

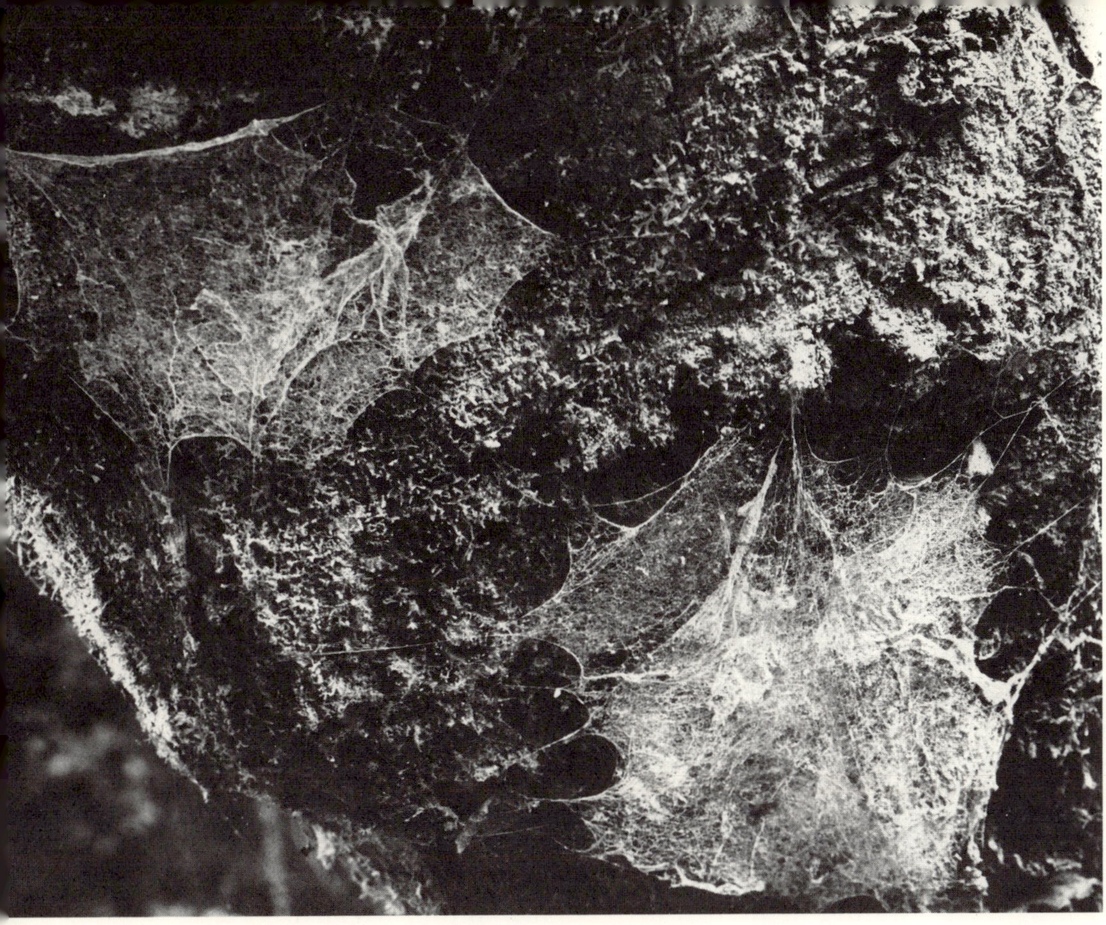

Fig. 148 Snares of *Ixeuticus dalmasi* on a tree trunk.

Fig. 149 *Ixeuticus dalmasi* female.

Fig. 150 The Sombrero Spider, *Stiphidion facetum.*

used by Professor V.V. Hickman seems an apt one which could well be used as a common name.

The commonest amaurobiid spiders in the forest belong to the genus *Maniho* (Figure 151). These agile spiders with brown carapace and legs and a dull, greyish abdomen, sometimes relieved by a double row of pale spots, construct a small sheetweb under loose bark, logs or in hollow trees. Nothing is known of their habits other than the form of the web and even the shape and structure of the egg-sac have not been recorded. However the use of cribellate silk plays little part in the life of these spiders and the cribellum is always small. It is, therefore, not surprising that the Agelenid genus *Amphinecta* seems to be the closest relative, differing mainly because these spiders have gone one stage further, having no trace of the cribellum or the associated calamistrum.

Fig. 151 *Maniho* sp. male.

Fig. 152 The delicate ladder snare of *Matachia*.

Some extremely delicate ladder webs found in the forest were recently the subject for study by Professor B.J. Marples who found that they belonged to two species of *Matachia* (Figures 152, 153). The favourite homes of these spiders are the holes in the galls produced by the moth *Morovia subfasciata* in the woody creeper *Muehlenbeckia*. Professor B.J. Marples recently published the results of many years study of these spiders and we can do little better than quote some of his observations. ' . . . a well developed web is constructed. This consists of a number of more or less parallel threads, ten centimetres or more in length, attached to the neighbouring twigs but not radiating from the hole in which the spider lives. A small network of threads joins them to the hole in which there is very little silken lining. All these threads are of plain silk. The calamistrated threads are laid down on the parallel threads in a ladder-like fashion. A calamistrated thread runs for a short distance

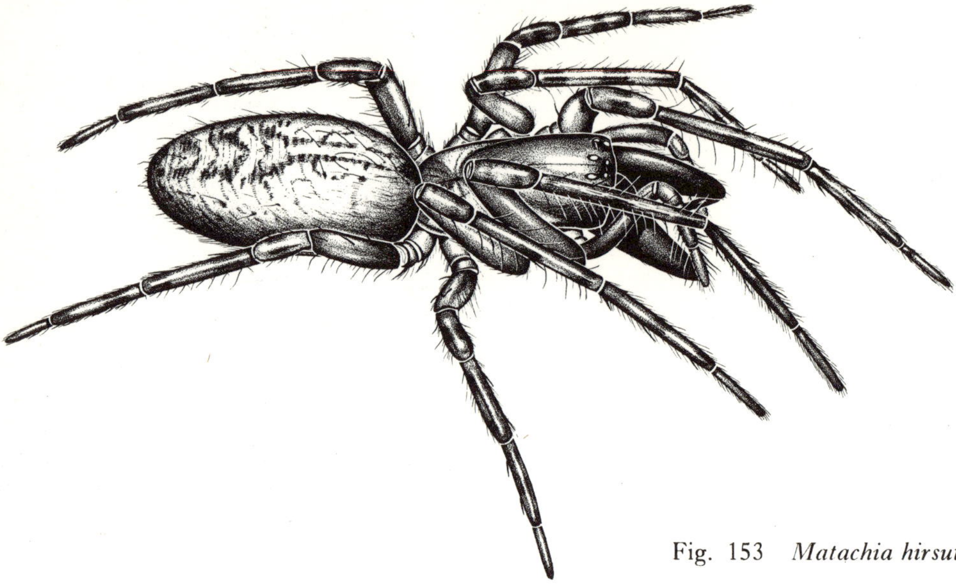

Fig. 153 *Matachia hirsuta* female.

along one of the plain threads, crosses to another parallel one, proceeds along it for a short distance and then back to the first, and so on, This ladder-like arrangement of the calamistrated threads is common in the Dictynidae. A fairly extensive web may be constructed in one night's spinning. It may be added to subsequently and remain in use for some time. Inside the hole, the spider rests with three pairs of legs pointing towards the opening and frequently the long anterior pair may be seen projecting and resting on the threads of the web. Remains of prey are seldom found in the web, but moths and beetles were observed. Captive specimens were observed to drag small insects into the hole and to eat the larger ones out on the web. If the twig with the hole is gathered the spider readily leaves it.'

The egg cocoon, which is laid in December and January, contains on the average only a few eggs, perhaps only a dozen or so, held together by a thin covering of white silk and attached to the wall of the burrow. As with many of the tunnel dwelling spiders, including the trapdoor spiders, after the young emerge from the egg-sac they will often continue to live in the burrow with the mother until they are all well grown and during this time presumably share with her the food trapped in the web. It also seems that when they wish take up an independent life they do not wander far but take up residence nearby. Colonies are found scattered throughout the forest and the spiders may be numerous in one patch but absent from the next. Because the adult male does not have a functional cribellum and therefore would not be able to construct an effective web, it is thought that he must die shortly after mating.

MEGADICTYNIDAE

In 1906, the German arachnologist F. Dahl, described a spider from New Zealand which he thought was a large dictynid. This rather dull-looking spider with a mottled brown abdomen and pale brown carapace marked with a black streak down the middle, he named *Megadictyna thileniusi* (Figure 154). Nothing more was heard about this spider until almost fifty years later when R.R. Marples found specimens in a collection he was studying. At this time, he was at the University of Otago preparing a thesis on dictynid spiders. He overlooked the obscure, and by then practically forgotten, description of Dahl and renamed the spider *Ihurakius forsteri.* One effect of the rebirth of this spider into scientific literature was that the attention of a Finnish specialist, Pekka Lehtinen, was drawn to this species while he was studying cribellate spiders of the world. After investigating further specimens, Lehtinen concluded that this rather insignificant-looking animal was in fact one of the most primitive of true spiders which he felt should be placed in a special family of its own.

Fig. 154 *Megadictyna thileniusi* female.

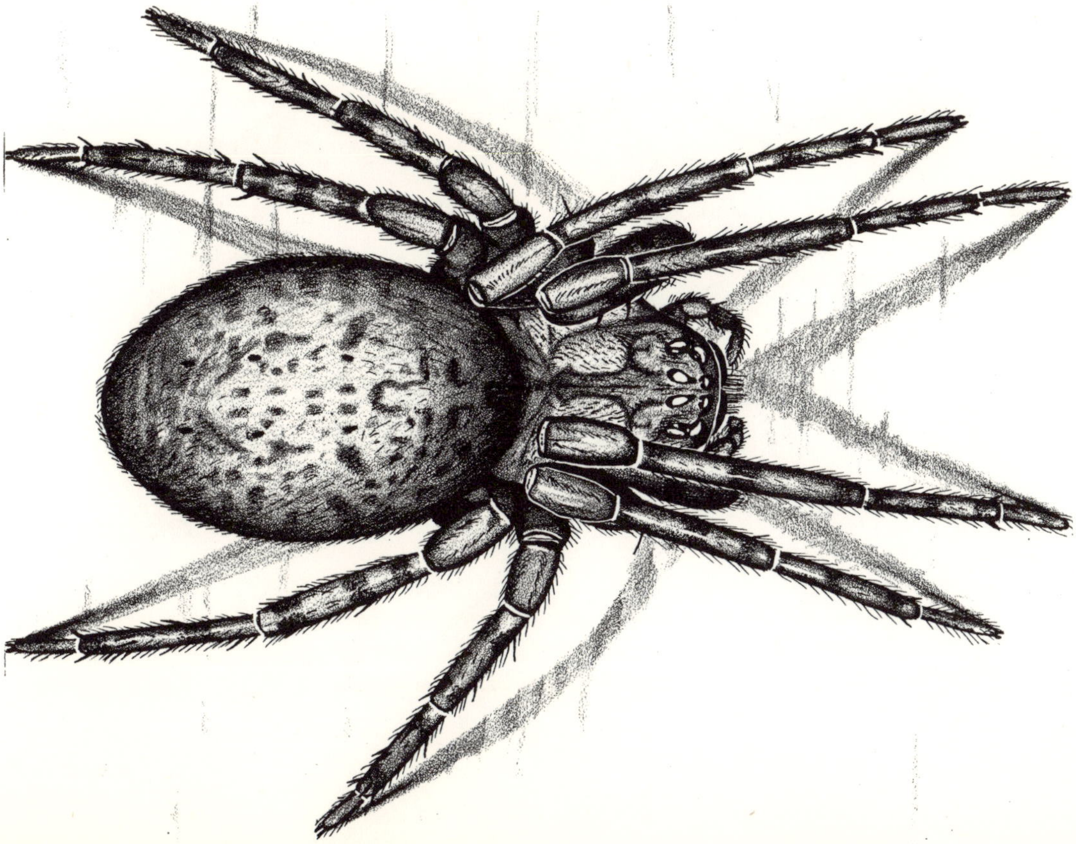

Because Dahl gave a name to the spider first, it is now known by his name of *Megadictyna thileniusi* and the family as Megadictynidae. As is so often the case, with the sudden burst into prominence of this spider, we find that we know very little about it. There are actually a number of species found in the forest of both North and South Islands. In spite of the well developed cribellum they make little use of cribellate silk but construct a very small sheetweb on or near the forest floor. The egg-sac is large and contains between two and three hundred eggs held together with a thin covering of silk and is suspended near the web. Apart from these meagre facts, nothing is known of the spider's habits.

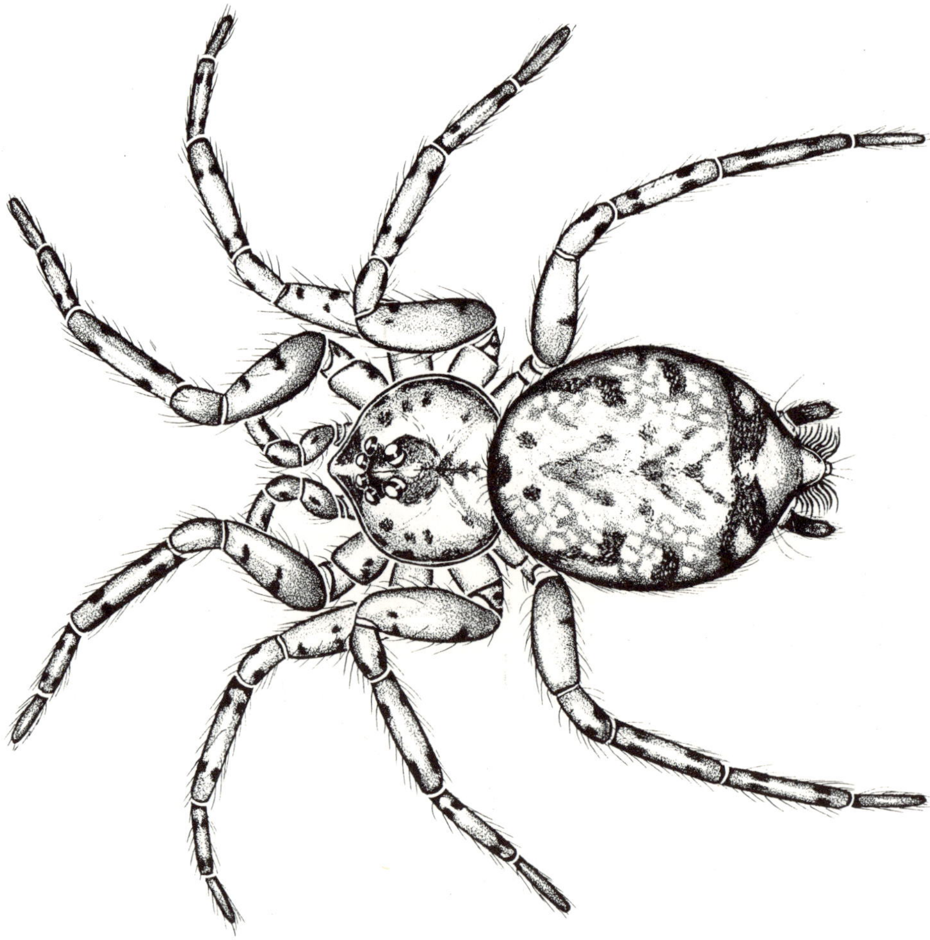

Fig. 155 *Oecobius annulipes* female.

OECOBIIDAE

The oecobiids, which are tiny spiders never more than a few millimetres long, have taken to living on the walls of houses and have now achieved a worldwide distribution. They weave a small sheetweb never more than an inch wide across cracks and corners beneath which they rest flat against the surface with their legs spread out to the side. It seems certain that there are no native species and that the one we find in New Zealand, as yet only in the North Island, is the common species of the northern hemisphere – *Oecobius annulipes* (Figure 155). This spider is pale, yellowish-brown with a few black patches which show clearly against the paler colouring. However, the feature which makes it distinctive along with other members of its family is the relatively huge anal tubercle above the spinnerets. This is decorated with two conspicuous rows of long hairs, although it must be admitted that the spider is so small that to see this structure clearly one needs to use a microscope.

ULOBORIDAE

The fact that spiders of this family are common in our forest escaped everyone until 1946 when G. Chamberlain described a spider he called *Uloborus waitakeriensis* from the bush near Auckland (Figure 156). Now we know that they are to be found in the forest in most parts of

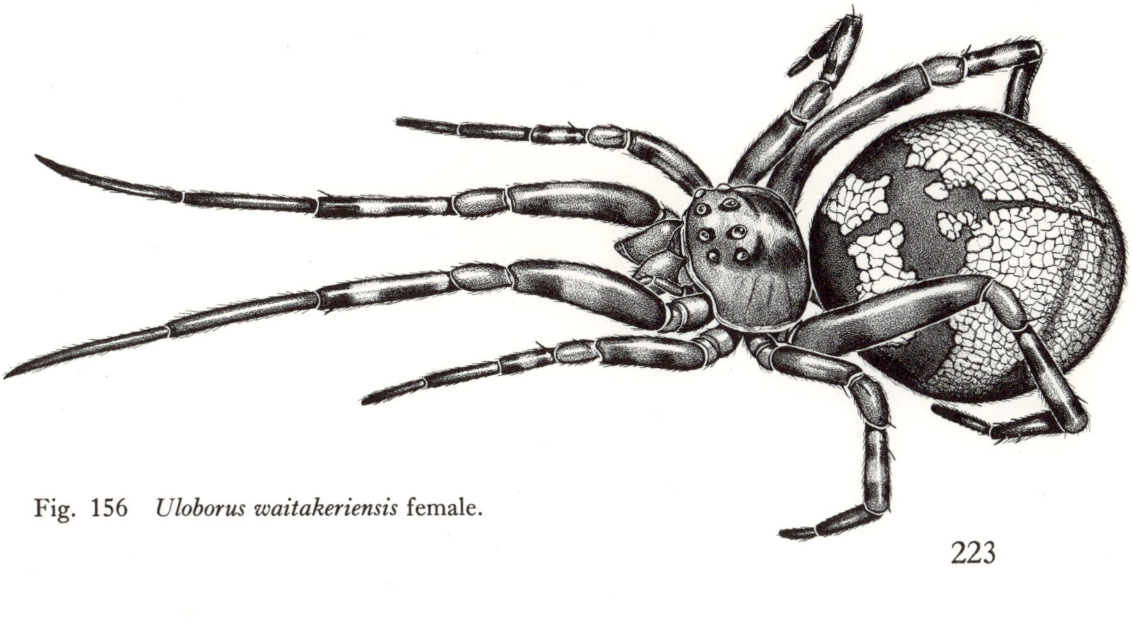

Fig. 156 *Uloborus waitakeriensis* female.

New Zealand. These small spiders which are about a quarter-of-an-inch long, construct an orbweb, differing little from the webs made by the ordinary orbweb spider except that the sticky spiral is spun in cribellate silk. These webs are often horizontal or distinctly inclined; the spider is always found at the hub. The abdomen is conspicuously white with a few black patches to the sides but the carapace has a pale oliver-greenish tinge, apart from the large black area behind the eyes, and the legs are also olive with dark bands.

If the observations made on overseas species are valid for ours, then the mating habits will be similar to those of the typical orb weavers. W.S. Bristowe reports on the English species that, when the female is ready for egglaying, she makes a rudimentary web, comprising a hub, a few radii and perhaps a few spirals. From this she hangs a few irregular elongate, lightish-brown egg-sacs of papery silk which are abandoned after a few days.

125. *Uloborus waitakeriensis*. Female. Family Uloboridae. Pohangina. Length 3 mm.

126. *Ixeuticus martius*. Female. Family Amaurobiidae. Nelson. Length 14 mm.

127. *Oramia* sp. Female with egg-sac. Family Amaurobiidae. Dunedin. Length 16 mm.

128. Undescribed amaurobiid. Family Amaurobiidae. Pigroot, Otago. Length 4 mm.

129. *Maniho tigris* sp. Female. Family Amaurobiidae. Kaituna Valley. Length 5 mm.

130. Webs of *Ixeuticus dalmasi*. Family Amaurobiidae. Pohangina.

131. *Ixeuticus dalmasi*. Female. Family Amaurobiidae. Mangareia. Length 7 mm.

132. *Matachia hirsuta*. Female. Family Amaurobiidae. Dunedin. Length 7 mm.

CHAPTER SIXTEEN

Poisonous Spiders

To many people the mere mention of a spider conjures up pictures of spider bites and their dire effects, leading to a conviction that they are animals to fear and shun. No doubt the fact that these animals live by violence and the use of poison, builds up a sense of repugnance but, once poison is accepted as part of their way of life, an appreciation of the multiplicity of their behaviour patterns and the intricacy of their actions is developed.

Tales of the effects of spider poison on humans are legion but not all are based on fact. There are probably no more than a few score of species throughout the world which cause more than short-term discomfort to the person they bite. The classical case of so-called spider poisoning is associated with the lively dance of Italy, called the tarantella. In the vicinity of Tarento in Italy during the Seventeenth Century, a peak was reached in a craze which had been known spasmodically for some centuries before in which a person said to have been bitten by the common wolf spider of the area, *Lycosa tarantula*, danced wildly hour after hour to the accompaniment of music until exhausted and in this way worked out the poison from his system. This phenomenon became so widespread at that time that special music was prepared and during the harvest periods roving fiddlers would move from field to field waiting to be hired when someone was bitten. Some of our readers might conclude when viewing the present day trends in music that perhaps a spider bite was not really necessary to produce this craze. As it happens, this is the general view today. Numerous experiments have been carried out to test the toxicity of the venom of this wolf spider, even to the extent of having specimens bite volunteers. The comments of these researchers have ranged from 'causes less inconvenience than a gnat bite' to 'the effects were insignificant' and there is no reason to believe that the toxicity of this spider is any different today from what it was some centuries ago.

In general, it has been found that wolf spiders are harmless but this generalization does not hold for all because the bite of a few including the notorious *Lycosa raptoria* of South America, may cause extensive local lesions, but none seem to produce the overall effects which might account for the dancing mania. Many theories have been put forward to explain this extraordinary epidemic, from those suggesting that was part of tourist promotion at those times, to one which suggested that all the dancers were adherents to the ancient cult of Bacchus, who used the spider bite explanation to conceal the real reason for their behaviour. A few of the recorded cases at that time were probably based on spider bites but the symptoms of these suggest bites by the local species of *Latrodectus*, spiders related to the New Zealand katipo.

While the wolf spider which was blamed for the dancing epidemics of Europe is called the Tarantula, to most people this name belongs to the large mygalomorph spiders which abound in the tropics, the so-called bird-eating spiders, and this term is often extended to include the trapdoor spiders and their kin. Despite their large size and fearsome appearance, very few of these mygalomorph spiders have been indicted as being dangerous to man. A number of cases of bite from the tunnelweb spiders *Porrhothele* and *Hexathele* has been noted in New Zealand but the effects have been slight. Sometimes nothing more has happened than one would expect from a puncture of the skin, while in other cases the bite has resulted in some slight swelling and occasionally a raised blister. Only one case lists any serious symptoms and it was possible that these were due to a secondary infection of the bite. The story is drastically different with the related tunnelweb spiders of New South Wales which belong in the genus *Atrax*. A number of deaths have been attributed to these aggressive spiders, often with the death of the victim occurring within a few hours of being bitten. Fortunately, none of these spiders occur in New Zealand although our common tunnelweb spider *Porrhothele antipodiana* may alarm Australian visitors because its appearance and habits are so similar to their own spiders.

It is probable that many of our spiders could cause some discomfort if they were able to bite but there are fortunately many factors which make this possibility remote. Some are simply not big enough for their fangs to penetrate the skin but many which should be physically capable of biting do not do so because a certain sequence of actions must take place first. This applies to many of the hunting spiders as well as web spiders.

Most of the spider bites, other than by the katipo, which have been recorded in New Zealand have occurred when the spiders have been

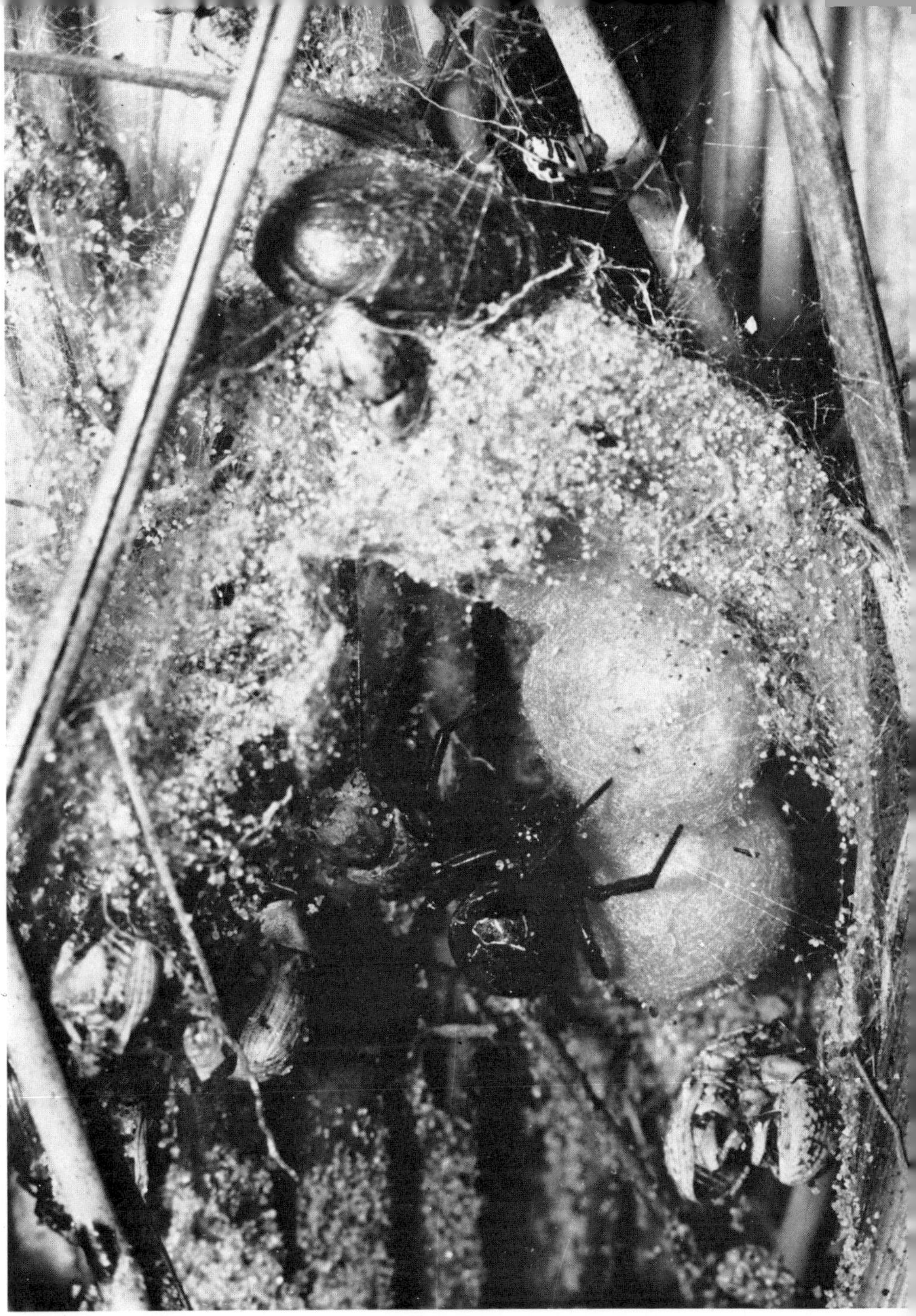

Fig. 157 The katipo in her snare. Note the larder of beetles and the young spiderling above the web.

entangled in clothing. A number of these have been orbweb spiders and perhaps by coincidence both of the species which have been actually identified have been widespread Australian species introduced into this country. Of the five bites which have been recorded, four by the common *Aranea pustulosa* resulted in local swelling with the development in an hour or so of a small blister. The slight pain in the region of the bite disappeared within a few hours. The culprit in the fifth case was the large *Aranea transmarina* usually known in New Zealand as *A. brouni*. An elderly lady was bitten on the finger by a female of this species and suffered considerable pain which spread over the hand and arm; in this case the pain remained in the joints for some days.

Chamberlain (1947) tells of a person bitten by the large sheetweb spider of the North Island, *Cambridgea foliata*.

> The punctures were in the left thigh about midway between the hip and knee joints. Only slight pain was felt at the bite and no symptoms appeared for about six hours. At this time an intense itching developed, and a swelling commenced at the site of the punctures. At the end of the third day the swelling covered an area roughly seven inches in diameter, and was causing considerable pain. Vesication was observed at the site of the punctures, consisting of a lymph-filled blister about half-an-inch in diameter, and surrounded by a narrow erythematous area. On the fourth day the swelling began to subside. A scar tissue formed under the blister, and the upper layer of skin was sloughed. At the end of the week the condition of the patient had returned to normal."

There are numerous species of *Cambridgea*, many large with huge fangs, which build their sheet webs in the hedges around houses as well as in outbuildings and these are often found in the house. If these spiders are, in fact, capable of inflicting such drastic bites it is surprising that so little is heard of them. Our own experience of them is that they will strike only at insects caught in their webs and none of the hundreds we have handled has shown any inclination to bite.

Without attempting to minimize the possible effects of spider bites in New Zealand, it can be said, in general, that only one species is to be in any way feared. This is the katipo which, although so well known by repute, is not so well known when it comes to recognition of the spider itself. Many are surprised to find that the katipo is quite small. The abdomen of a fully grown female is not quite as large as a garden pea, round and shiny and usually black with a bright red stripe down the top surface and sometimes with a red patch underneath. It is very distinctive and easily identified by the bright red stripe but unfortunately as far back as the 1880's a colony was found off Mahia Peninsula in Hawkes Bay which had no stripe and so were entirely black. As it happens, throughout New Zealand there are a number of species of the related genus *Steatoda* of about the same size and form, also characterised by their black or dark-coloured bodies, including the most common of the introduced house spiders. As can be imagined, much confusion and unnecessary fear has been generated by these relatively harmless species in and around the house. In fact in the eighty years since its discovery no further colonies of the black katipo of Mahia have been found, and the many reports of the black katipo are all based on mistaken identifications. It has been suggested that the so-called black katipo from North

Fig. 158 The katipo flings threads with her hind legs to entangle a Rove Beetle.

Auckland, which is an unidentified species of *Steatoda*, may produce some ill effects, but, although this may be so, no definite cases have been recorded.

The katipo belongs in the widespread genus *Latrodectus* along with many other species which have been implicated in serious cases of poisoning, some leading to death, at one time or another. These spiders are consequently one of the few groups which have their own common, if not 'popular', names wherever they are found. The northern hemisphere species *L. tredecemguttatus* is called the *malmignatte* in Italy and France, while it is the *Karakuri* or 'black wolf' in Russia and is known by many other names in the other areas in which it lives. In South Africa *L. indistinctus* is known as the 'shoebutton spider' or *knoppiespinnekop* and another species is called the *menavodi* in Madagascar and so it is possible to list some scores of names which highlight the fact that the poisonous propensities of these spiders rarely go unnoticed. The best known name is undoubtedly the 'black widow' in North America for the common species of that region *L. mactans*. The name black widow refers to the popularly held, but erroneous, belief, that the female invariably feasts upon the male on the completion of his nuptial duties.

Dr H.W. Levi of Harvard University who recently revised the genus *Latrodectus* on a worldwide basis came to the conclusion that the New Zealand katipo was the same as the Australian Red-back and that this same spider was distributed widely over the western Pacific area and extended as far as India. He further decided that it was a form of the true black widow spider of America, which would mean that the accepted name for this form would then be *Latrodectus mactans hasselti*. However, while this solution tidies up the mass of scientific names which have been used previously, it is probable that the true facts of the relationships of these spiders have not been settled. The Red-back spider occurs over much of Australia and Tasmania, where it lives in warm dry areas and even in open bush and may also be found in gardens under loose stones and in piles of rubbish. The katipo of New Zealand, however, occurs in a very restricted habitat. It is found only near the sea and then usually where the beaches are sandy. The tangle webs are always constructed close to the ground, mostly at the base of tufts of Marram grass or sedges, but also under driftwood and, if they are available, under rusty tins lying on the sand. It is this last man-made habitat which they seem to prefer and which is the first to be examined if we want to find katipos quickly. With very few exceptions, they are rarely found more than a few hundred yards from the sea, and these exceptions are seen only when sand-dunes extend back from the beach.

Fig. 159 Male katipo one moult from maturity.

The striking differences shown in habitat preference between the spiders Australia and New Zealand together with slight structural differences may indicate that they are not really the same, and for this reason we retain for the present the name *Latrodectus katipo* for the New Zealand examples.

The restricted occurrence of the katipo explains to a certain extent the paucity of recorded bites, although its timidity is also a contributing factor. The katipo is found along most of the coast of the North Island and as far south as Dunedin on the east coast of the South Island and probably well down the northern half of the west coast, although the records from this area are few. This distribution means that the preference of the katipo for sandy beaches and warmer climes coincides with that of the majority of surfers and bathers, both being present in large numbers along our beaches during the summer. Recently when collecting these spiders near Christchurch, we had to wend our way through sunbathers stretched out on the sand while gathering specimens which had constructed their webs at the base of nearly every tuft of Marram grass, a concentration of from three to four webs for every square yard. If this virtual contact between spiders and man is multiplied to cover the whole coast, and it is realised that there are very few cases of katipo bite, it becomes clear that the chances of being a victim are very remote; there would be even less chance if everyone knew something of the appearance and habits of our only dangerous spider. Only the female katipo is capable of biting and this only when it is adult and hence most conspicuous, with its black colour relieved by a prominent red stripe. The male is much smaller than his mate and seems completely different. The carapace is brown with a central black band and the abdomen glistening white with three black streaks on the side and two irregular black bands down the top enclosing a string of orange diamond shaped patches (Figure 159). The adult male does not differ much from the immature stages of both sexes; one way of looking at it would be to say that the male never really grows up. Apart from their size, the only difference which can be seen in the colouring of the spiderling as it emerges from the egg-sac is that the pair of black bands down the back are at this stage represented by six pairs of black patches and the median diamonds are pale yellow. By the time the female is one moult from maturity, she is mainly black and the dorsal stripe has intensified to red but she still has traces of the white markings down the side of her abdomen which will usually disappear after her final moult.

The katipo belongs to the family Theridiidae and like most of her relatives builds a cobweb – a tangle of threads of which only a few are sticky. Many of the species of *Latrodectus* found in other countries build a retreat in which the spider remains until something is caught in the web but the katipo is usually found in the main body of her web. The web is always near the ground so that the main catching threads are just above the surface; the bulk of the prey caught are beetles. Many of these beetles are much larger than the spider but once entangled they seldom escape, not because the poison is so deadly but because of the methods she uses to make sure they are firmly enswathed.

At the first movement of the trapped insect, the spider jerks and pulls at the threads probably not to avoid entangling the insect further but to find a thread which by resisting in a similar way a fishing line indicates the direction of the victim. She then moves down to the struggling insect but, instead of plunging in her fangs, she will stop short and turn around so that the spinnerets on the end of her abdomen are pointing at it. As with most of the theridiids, the tarsi of the hind legs of the katipo have a row of strong curved bristles which are arranged as a comb and she uses these to scoop sticky silk from her spinnerets and fling over the insect with a series of rapid movements which look just like the kick of a horse (Figure 158). After the insect is entangled further, she then runs up and down, to and from, the main portion of her web, laying down silk as she goes strengthening the attachment to her web. She continues thus until the struggles lessen and her prey is firmly enswathed and it is now that she moves down to administer the bite which will end the encounter. Once the insect has been poisoned, it is hauled up well above the ground to hang until she is ready to eat. While food is available she continues to trap, and so often her larder will contain a half-dozen or more insects awaiting her pleasure (Figure 157). Actually it is difficult to know whether an insect has been fed upon because these theridiid spiders feed differently from most others. Instead of crushing the insect as they exude their digestive fluids, they force the fluid into a small hole through which they also suck the digested body material so that when the meal is completed the original insect remains but is now only an empty husk. Normally this shell is hauled to the upper reaches of the web or cut away and dropped but sometimes it will remain in the larder.

It seems to be a widely held belief that once the male approaches the female to mate he is under sentence of death, but it is certain that he is seldom killed. The male stops and vibrates the web at intervals as he moves slowly in from the outer threads with the female at first lunging each time in his direction. If she continues to react violently as he comes near, he will turn tail and run, but if she remains passive he will reach forward with his front pair of legs and gently tap her as she hangs upside down from the threads. He then moves over her body until his abdomen is over hers and he is facing in the same direction. He inserts each palp alternately a number of times. There seems little haste and the mating may be spread over half an hour. After it is completed the male moves leisurely away resting at the margin of the web where he usually begins to run his palps through his fangs and then later recharges his palpal organ.

The egg-sacs appear during November and December. They are round, slightly less than half-an-inch in diameter, and cream in colour and each female constructs at least two and sometimes three or four in rapid succession. She hangs them up in a bunch in the body of the web and then spins more silk above, often loaded with drifting grains of sand to form a conspicuous canopy. At this period she is normally found clinging upside down beneath the sacs.

The spiderlings which emerge from the end of January always move upwards to the tops of the clumps of grass and then drift around in the breeze. With this tendency to balloon, there seems no reason for them to be restricted in their habitat; consequently it is thought that there must be some factor which makes them dependent on the conditions prevailing near the beaches where they occur. The spiderlings grow slowly but reach maturity in the following spring.

THE EFFECTS OF KATIPO POISONING

The literature on the effects of *Latrodectus* poisoning is vast and some general comments on the group as a whole will probably be of more interest than a detailed account of specific cases. The initial pain usually occurs within half-an-hour of the bite and the general symptoms become evident within a few hours, but, of course, the effects of the poison may vary with individuals and the amount of venom injected. In a typical case where a person has been bitten on one of the extremities, the pain will progress to the trunk of the body, resulting in considerable pain in the abdominal region which then becomes rigid and boardlike. What may follow varies considerably but the following is a quotation from the summation of possible symptoms collated by Thorpe and Woodson in their book on the black widow; fortunately not all these symptoms are expected to appear in each case.

'There may be various other symptoms including cold perspiration, an increase in blood pressure, a mild rise in body temperature, nausea and anorexia, vomiting, slight twitching or spasms of the muscles of the extremities, malaise, priapism, urinary retention, constipation, speech defect, insomnia, local edema, amnesia, difficult breathing, restlessness, cyanosis, depression, vertigo, chills, prostration or shock, paralysis, a macular skin eruption, jaundice, convulsion, tremors, muscle twitching, anxiety, and an increase in the polymorphonuclear leucocytes and delirium'. The poison particularly affects the nervous system and because of the rapid dispersal throughout the body the host of specific remedies in the past have rarely proved of much use. Comparatively recent developments have, however, greatly improved the chances of rapid relief to those unfortunate victims. The Commonwealth Serum Laboratories of Australia have developed an antivenom from the poison of the Red-back spider and this is now available in New Zealand for treatment of the katipo bite. According to Dr R.V. Southcott, this may prove beneficial even when given eighty hours after the bite.

The presence of the katipo was recorded as long ago as 1855 by T.S. Ralph who published a paper "On the Katipo. A supposed poisonous spider of New Zealand" in the *Journal of the Linnaean Society of London* for that year. In fact, the spider and its poisonous capabilities were well known to the Maoris who gave it the name 'katipo' which can be translated 'night stinger', distinguishing it from all other spiders, which were grouped together and called 'Punga-were-were'. It is interesting to read in the account by an early missionary, the Revd. Chapman, that 'the natives generally avoid sleeping on the sea beach, but have no fear of the Katipo half a stone's throw inland from the beach line', because it is clear that the Maoris also used the name katipo for the shiny black species of *Steatoda* which even today takes the place of the katipo further away from the beach and is generally harmless.

In a land free from harmful animals, it is not surprising that much attention has been given to its sole poisonous creature, and we find that some of the original papers presented to the New Zealand Institute after it was founded as the first major scientific body in the country discussed this spider. In the second volume of its transactions published in 1869, Dr F.W. Wright reported on four cases in the Auckland district, some going back to the earliest days of colonization. Since then, numerous other accounts have appeared in journals both in New Zealand and overseas, covering some forty cases. In spite of the often outlandish treatments prescribed in the past, the great majority of these unfortunate victims recovered after a few days although the effects were often severe.

CHAPTER SEVENTEEN

How to Find and Study Spiders

Spiders are everywhere where there is food to support them, but in New Zealand this is not always evident because a very high proportion are nocturnal. So it is necessary to explain in some details where they may live, how they may be found, and then studied.

At first, you will want to be able to identify the animals you find. Probably the best way to do this is to make a collection of preserved spiders which can be examined in detail under a magnifying glass or microscope. As you become familiar with them, you will find that most can be identified in life by small characteristics which you note almost unconsciously – the way they move, the webs they make, their colour patterns or their reactions when disturbed. These are things which cannot easily be written down, but will come to be detected with experience.

While it is most rewarding to have these animals under observation in their natural habitats, this cannot be done if the spiders come from far afield and techniques must be developed to keep them alive and happy in captivity so that they can be carefully watched while the significant features of their life-histories and behaviour are noted. You might find an unknown egg-sac and wish to rear the spiderlings to maturity or perhaps observe the mating behaviour and egg-sac construction of some particular species. It is much easier to carry our the preliminary observations at home, particularly as much of the activity will be at night. Then armed with this knowledge, you will be able to pick the right time to check your findings by watching them in their natural environment.

COLLECTING METHODS

When setting out to know the fauna of a particular area it is usual to use a number of mass collecting methods which very quickly produce a large varied collection which can be preserved, sorted out and identified. These are mainly methods which will tell you little more than the general types of habitats that the particular spiders prefer, but it is essential to know these facts before you are able to go much further.

The most productive method is beating. To do this we use one of the light-weight plastic dishes normally sold for developing photographic prints. The dish is held under a shrub or clump of grass while the foliage is shaken vigorously or beaten with a short but stout stick. Most of the spiders resting on stems or leaves will drop into the dish on the end of a thread of silk and can then be captured, but, if time is short, the whole collection, debris and all, can be shaken into a plastic bag and brought home to check through at leisure. The spiders rarely attack each other amongst the debris, but be sure that the bag is not placed in sunlight, otherwise the heat will quickly build up inside and everything will be cooked.

A similar, and frequently quicker method often used to catch grass and tussock dwellers is sweeping. An ordinary butterfly net made of stout material reinforced around the rim is swept through the foliage, catching everything as it gets knocked from its perch by the sweep of the net. It is simpler to empty the contents of the net into a dish before attempting to gather the animals which have been caught, but alternatively, if the net is held with the bag down, the spiders will crawl up the sides where they can easily be captured.

These two methods will catch most of the spiders which live above the surface of the ground, but there are many which live in vacated insect burrows in the twigs and branches of trees and shrubs and the only way to gather these is to look for them specially. This would be difficult of every hole had to be examined, but fortunately all the spiders use silk in some way, even if only as a dragline. Consequently it is only necessary to find a hole where there is silk and then to split it open carefully.

A large proportion of our forest species live on the forest floor among the leaf litter and moss. To capture these, two main methods are used. The first is sieving, which is simply shaking the animals contained in a sample of leaf litter through a sieve of quarter inch mesh on to a dish or a sheet of white paper. A lot of fine debris will come through the sieve as well as the spiders and other forms of animal life. These animals tend to lie quite motionless, making it practically impossible to see them. To get over this difficulty, we usually bring the litter sample home and sieve it on to a white dish under a strong light so that in a very short time they come to life and scurry away to find darkness.

There is a simple apparatus called the Berlese Funnel which utilizes the fact that these animals dislike light and also that they are so dependent on high humidity that they will move from a dry to a moister place. The apparatus we use is made from a four-gallon kerosene tin open at the top and cut and soldered at the base to form a funnel (Figure 160). A square of gauze of quarter inch mesh is fitted inside the funnel to hold the leaf litter and a lid with an electric bulb fixed into it is placed on top. The light and the gradual drying out of the litter causes the animals to move down to the gauze and then drop down the funnel where they are caught in a dish below. All that is required after the sample is inserted is to sit back and wait.

In open country, particularly grassland, it is difficult to find the spiders which do not make webs and here a pitfall trap is most effective. The only requirement is either a tin with the top smoothly removed or a straight-sided glass jar. If this is sunk into the ground so that the opening is flush with the surface and left for a few days, a surprising number of the ground spiders will drop into the hole and not be able to crawl out again. If you want live spiders a few holes should be punched into the bottom of the tin to let the rain water seep out but if you need preserved specimens a small quantity of preservative can be poured into the container after it is set.

If you hunt for individual spiders during daylight, it must be kept in mind that the nocturnal species will be sheltering somewhere in the dark. So the places to look are under bark, or logs and stones or at the base of shrubs. Many of the web builders will have secreted themselves near their web, either in a special retreat or at the end of a thread of silk leading from the web. A most useful way to bring out web spiders is to pretend to be a fly; the simplest way to do this is to carry a tuning fork with which to touch the web while the fork is vibrating. When using this method to catch spiders which come out from a retreat, it is prudent to carry a small stick with which the retreat should be blocked once the spider emerges.

Perhaps the most exciting and rewarding collecting technique is the use of a light at night. It is during these excursions that you are most likely to see spiders behaving in a natural manner. The best form of light is a lamp powered by torch batteries, which fits on your forehead. These lights have the great advantages of not dazzling the wearer and, as they throw a beam of light in line with your vision, you can then catch the glow back from the reflecting layer in the eyes of many spiders. This means that the spider's eyes show up as minute green lights which can be seen many yards away, long before you can actually see the spider itself. This phenomenon is particularly useful when searching for hunting spiders in open country. In addition, webs show up white and are more easily seen than in daylight.

HOW TO USE COLLECTING TECHNIQUES IN PARTICULAR SITUATIONS

Having described the various ways by which spiders may be collected it may be useful to explain how these methods can be applied to particular habitats.

FOREST

The greater proportion of forest dwellers are nocturnal and, in order to get some idea of the wealth of species which are present in even the smallest patch of bush, the best introduction is an hour to two at night with a light. Most spiders become active some time after night fall, in the summer around 10 p.m., and the peak activity is reached by midnight. It is during this time that the web builders begin to repair their webs or begin the construction of new ones and the hunting spiders actively seek food. The first thing you will notice is the mass of silk gleaming from every shrub and fern, but as many of these threads are draglines left behind during hunting forays or while drifting from branch to branch, a search for an associated spider will often be in vain. However, a close look at some of the single threads suspended between twigs will soon show that at least some of these are the simple snares of minute spiders which use the thread as a kind of aerial tripline and the spider will be found clinging on the thread or at the base. The owners of the larger snares will usually have moved out from their retreats and be waiting for victims to become entangled, but if you want to catch these spiders you will need to carry a small stick in order to block the retreat before you try and capture the spider itself. The hunting spiders will be seen running about or resting on the trunks of trees, the leaves of shrubs or the forest floor. The tunnel dwellers, such as *Ariadna* or the forest trapdoor spiders, will be crouching at the opening of their retreats waiting to pounce upon any unsuspecting insects which come close enough. This is also the time you will find the ungainly long-legged harvestmen striding over the shrubs or the trunks of trees as they search for food.

To find these same animals during daylight, you will need to beat the low shrubs and ferns, turn over logs and stones, sieve the litter and moss from the forest floor or carefully trace back from the webs along the signal thread to the spider's retreat. A pitfall trap in the forest floor provides much information on the movements of the ground dwellers, but will also capture the males of many web builders while they are wandering in search of a mate.

GRASSLAND

While sweeping will bring to light many of the spiders which construct their snares among blades of grass, the only way to be sure of locating some of the vagrant spiders is to set out pitfall traps. Along with many of the hunting spiders, the males of the various trapdoor spiders will also fall into these traps, but not the females who remain in their burrows. To find the female trapdoor spiders a sharp eye is needed to distinguish the outline of the lid or small pile of excavated soil and then it is necessary to dig out the spider from the ground.

RIVER-BEDS

It may be thought that river-beds, continually subjected to sudden flooding, would be barren but, in fact, shingle river-beds in New Zealand have a varied fauna, seemingly derived from the sub-alpine region, which is peculiar to this habitat alone. Most of the river-bed spiders hunt at night and may be seen with a light, but the most effective though rather strenuous way of finding them is to turn over the loose stones under which they shelter during the day. Some species will only be found close to the water, but generally they are concentrated on the higher areas of the river-bed which are not subjected to continual flooding. A pitfall trap is also effective so long as the river does not flood.

SEASHORE

A surprising number of species are found only near the sea, and many of these in very restricted regions. The sandy area above high tide level is the home of the katipo and also the pale sand wolf spiders, while the same region above a pebbly beach supports the black beach jumping spiders and a number of dark coloured wolf spiders. The wolf spiders are best lookedfor at night with a light, but a pitfall trap is also effective. A number of spiders live only among the piles of stones which accumulate above the high tide level and the only effective way to find these is to clear a face and slowly work into the pile by removing each stone carefully. If there are bare rock faces within the splash zone, various species of the trapdoor spider *Migas*, the tunnel dwellers *Amaurobioides* or the six-eyed spider *Ariadna* will live in their retreats attached to the rock or perhaps built into cracks.

There is only one species which can in the true sense be called a marine spider. This is *Desis marina*, a most conspicuous reddish coloured spider which lives in the area between high and low tide levels, and to find it you will need to examine the cracks and crevices exposed when the tide is out or turn over the loose stones.

BUILDING UP AND MAINTAINING A COLLECTION

Unlike most insects, it is not possible to make a collection of dried spiders which will have any scientific value or use, and so specimens are always preserved in fluid. Most museums use seventy percent or eighty percent ethyl alcohol for preservation sometimes mixed with a little glycerine to keep the spiders soft and protect them if the fluid evaporates. A few drops of acetic acid are often added because this is said to help preserve the colours. Methylated spirit may be used, but never formalin because this makes the specimens very brittle and difficult to study. As they become brittle, they are also likely to lose many of their finer and more delicate structures which are important for identification.

The preserved spiders are kept in the small glass vials which are available from chemists' suppliers and although in the past it has been customary to plug each of these vials with cotton-wool before placing them in a jar filled with alcohol for permanent storage it is now more common to substitute one of the many forms of plastic plugs which are readily available (Figure 161). Only specimens of a single species which have been collected at the same time and place are kept in each vial, together with a label giving the basic information pertaining to the specimens and the scientific name, if known. The more essential information which should go with the specimen is shown on the sample label illustrated (Figure 162). Notes should have been taken of any interesting and possibly important observations which were made while the specimen was alive, such as the form of the web, any unusual behaviour, ecological preferences and details of the colour of the specimen, but it will not be possible to include all this on the label you prepare. This information is related to the specimen by giving each notebook observation a number which is then included on the specimen label. Nevertheless, it cannot be stressed too strongly that complete

Fig. 161 Storing a spider collection.

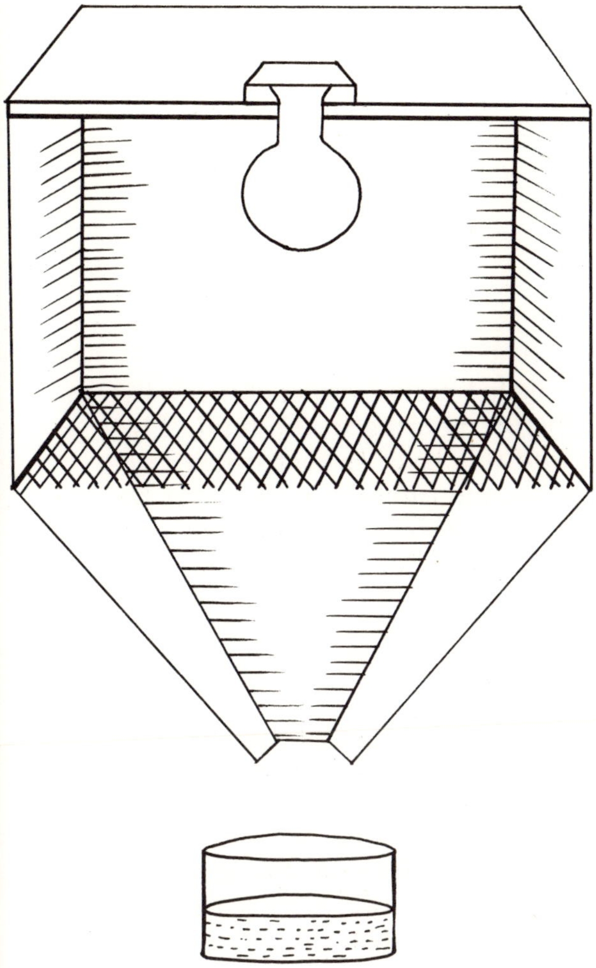

Fig. 160 Section through a Berlese Funnel.

reliance on a number system where only the number is included with the specimen is courting disaster, as the loss of your notebook will mean that all your efforts have been in vain and your collection becomes completely worthless.

We find that the standard pint preserving jars with replaceable lids are most satisfactory for storing the vials although almost any other jar with a tight fitting lid can be used as long as the lid is replaced if it corrodes.

Preserved spiders are examined when submerged in fluid, which is normally the fluid in which they are preserved. Spiders preserved in alcohol can be transferred to water for examination as long as there is not too sudden a change in the concentration of the fluid causing the spider to shrink. If the spider is preserved in 70% alcohol then a few minutes in 50% and then 30% before placing the specimen into water is all that is required. The most difficult problem when examining the different structures while the animal is submerged is to get the specimen resting the right way up, but there are a number of ways to achieve this.

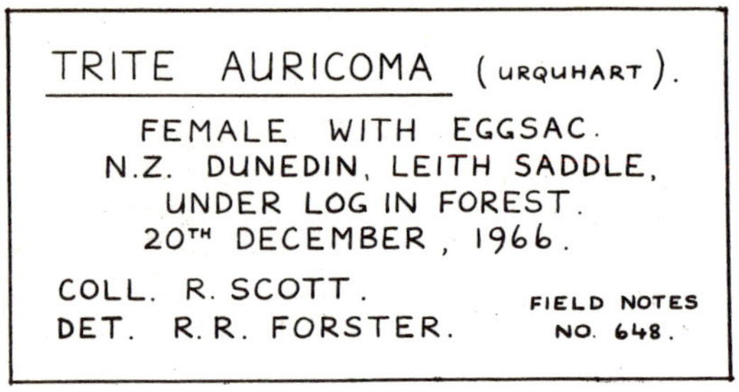
TRITE AURICOMA (URQUHART).
FEMALE WITH EGGSAC.
N.Z. DUNEDIN, LEITH SADDLE,
UNDER LOG IN FOREST.
20TH DECEMBER, 1966.
COLL. R. SCOTT.
DET. R.R. FORSTER.
FIELD NOTES
NO. 648.

Fig. 162 A typical label.

The most effective method is to smear a little Vaseline or petroleum jelly on the bottom of the examining dish before pouring in the fluid. The specimen can then be lightly pressed into this layer so that it takes up the position you want. As long as this is done carefully it can later be lifted without the Vaseline sticking. However, some students use Plasticine or prop the spider up with small balls of paper or pieces of glass in preference to Vaseline.

KEEPING SPIDERS ALIVE

Once you become familiar with the different species of spiders and know how to recognize them in life, then you are ready to study some particular aspect of their lives. We would stress that first you will need to be sure that you do know them well because many spiders which look alike are in fact quite different species, with probably different behaviour so that if they are confused your observations will be worthless. While it is not difficult to rear a spider from a half-grown stage to maturity, it is not quite so easy to take them through from the early spiderling stage or to breed them from one generation to the next. Some spiders live happily together but many are cannibalistic, so each spider must be kept in its own container. Another problem which will need to be overcome is a regular and reliable source of live food for them. To deal with this problem first; the most satisfactory method is to set up a culture of the fruitfly *Drosophila* which is easily maintained throughout the year. Most of the larger libraries have books which give details on preparing the food for these cultures and wild fruitflies can be trapped to begin a culture by setting out an over-ripe banana. For the larger spiders, house-flies or blow-flies can be caught as they are wanted as, fortunately, most spiders will thrive on only one or two a week. To keep larger spiders alive, a simple box with a sliding glass front can be used, or even, at a pinch, a glass jar (Figure 163). It is helpful to use a small dish or tin of earth containing a twig or plant so that this can be moved in and out of the cage as required after the spider has used it to construct a snare. Most of the activities are more easily seen from the side rather than above and so it is best to have the glass on one side. As the spider will be fed mainly on flying insects, it is difficult to get these into the cage through the moveable front but this problem can be solved by boring a hole in the side through which a vial containing the insects can be pushed. If the vial is kept dark by holding your hand over it the insects will move towards the lighted end and then out into the cage. When the hole is not being used for feeding

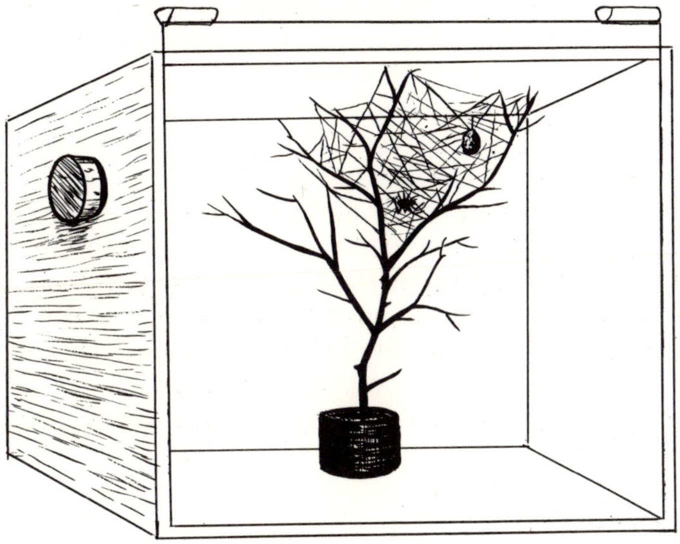

Fig. 163 Observation cage.

the spiders it is kept plugged with a cork. Many spiders will die unless the air is kept humid and this can be achieved by placing a damp wad of cotton-wool in the cage. Most spiders are able to drink any water they want from the moisture in the cotton-wool.

Some spiders normally live in old insect tunnels in the wood but these can be easily duplicated with the added advantage of making an artificial home in such a way that they can be examined when you wish. All that is needed is a small base of wood nailed to an upright section through which a hole is bored to take the neck of a small vial. A further three sides are nailed together so that they fit down over the stand and enclose the vial. Most of the tunnel dwellers will take happily to this improvised home, catching their food through the hole bored in the wood; if you want to examine them all you need do is to lift off the three-sided section.

With care and close attention to particular requirements for food and water, it is comparatively easy to rear the larger spiders in ordinary breeding cages and if the male is introduced at the right time mating will take place and eventually the egg-sacs be laid. The most difficult period of the spider's life is that which spans the first two or three moults after the spiderlings leave the egg-sac. The most successful method used so far to rear these spiderlings in large numbers is to place each one in a small section of glass or clear plastic tubing, plugged at each end with cotton-wool. The tubes may then be stored on racks, which if placed in glass dishes can be stacked one on top of the other (Figure 164). A piece of damp cotton-wool in each dish controls the humidity. As they become larger, most of the spiderlings have no difficulty in catching and eating the fruit-flies, but to get very young spiderlings to eat, it is often necessary to slightly crush a fly and leave it in the tube for them to stumble across. Moulting, even in small tubes, usually presents no difficulties but it has been found that some of the more active jumping spiders which always leap upon their prey must have enough room in their container to complete a jump. While it is usually necessary to maintain a relatively high humidity in the containers holding spiderlings, particularly the forest species, care must be taken to see that moisture does not gather on the inside or the spiderlings will be trapped by the surface film and die.

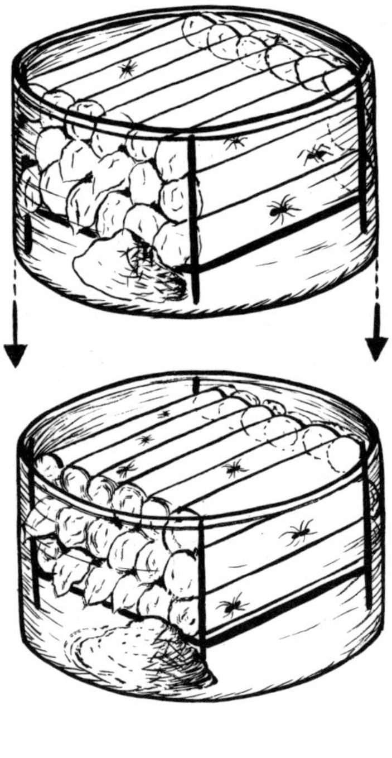

Fig. 164 A method for rearing small spiders.

PHOTOGRAPHY

Photographing spiders can be fun but it is also a most useful way to build up permanent records to supplement your notes. The photographs illustrating this book were all taken with a 35 mm single lens reflex camera while most of the black-and-white drawings are based on colour slides, which have been projected on to drawing paper to provide the original outline. Many texts on natural history photography suggest a prior sojourn in the refrigerator or perhaps a mild dose of an anaesthetic to keep the animal quiet while it is being photographed but a little patience, combined with a few less drastic techniques, will produce satisfactory results.

The main difficulty in obtaining good photographs is ensuring that there is a sufficient depth of sharp focus to show the features you wish to highlight. For this reason it is more practical to use electronic flash equipment, rather than to depend on available light. If this is available, the extension cord from the camera to the flash should be long enough for the flash to be used while not fixed to the camera. It should be standard practice for all photographs to be taken with the aperture of the lens fully closed, preferably to f.22, then correct exposure is dependent entirely on the distance the flash is held from the subject. Close-up photographs are made by extending the lens away from the film in the camera by means of a series of simple tubes or with a bellows attachment and sometimes in extreme close-ups with both. As the lens is moved away from the film by the tubes and bellows, less light will reach the film and so the intensity of the light on the subject must be increased to compensate for this loss. This is done by bringing the flash relatively closer. The distance the flash is held from the subject can be calculated with each photograph but, as this differs according to the extension used, often by the time you have worked out the details the subject will have gone. Actually, the distance the flash should be held rapidly becomes second nature if first one or two films are partly sacrificed while a series of photographs is taken under different conditions. It is essential at this stage to take detailed notes of the distance at which the flash is held and to have both dark and light backgrounds for the animals photographed. The reason for different backgrounds is that the light bounced back from a pale background will make very considerable differences to the exposure; whereas an animal on a dark background will be perfectly exposed, on a pale background it could be grossly over exposed.

If the spider is sedentary, then the camera can be placed on a tripod, but if it is moving it may be necessary to stalk it with the camera held in the hand. One way of securing pictures of spiders which will not stay still is to focus the camera on a fixed spot and wait until the spider moves over the spot before the shutter is released. Usually a little judicious shepherding will bring it into the position you desire. Some spiders seem always on the move and, with these, the knowledge that as a group they do not have a very efficient breathing system can be applied. If they are kept continually on the move for a couple of minutes they will then require a rest giving you plenty of time to bring the camera up and take the photograph you want.

Still photographs of moving spiders will often appear unnatural because this is not the way one usually sees them and so if the aim is to produce photographs to be admired rather than to record action it is best to concentrate on spiders at rest. While photographing spiders in their natural habitats, it is difficult to avoid including blurred and distracting pieces of the surrounding foliage. This can often be overcome by placing small sheets of card behind the subject as it is photographed, but care must be taken to see that the artificial background is far enough away to prevent shadows being recorded on the film.

Bibliography

We have listed below a number of publications which will be useful for those readers who wish to extend their knowledge. We have not listed individual papers because details of these can be obtained from the two world catalogues by Bonnet and Roewer which, as well as cataloguing all the species described throughout the world up till the time of publication, give a list of papers on spiders published in the various scientific journals. The *Systematic Catalogue of New Zealand Spiders* by A.W. Parrott provides a list of all New Zealand spiders described up to the date of publication (1946). Part 1 of *The Spiders of New Zealand* is a general introduction to the spiders, which includes a full list of the scientific papers relating to our spiders. This latter publication of which two parts have been published will eventually describe the complete New Zealand fauna.

The key to the identification of harvestmen published in *Tuatara* is useful for the preliminary identification of these animals but for more detailed information on the Laniatores, the *Canterbury Museum Bulletin* No. 2 should be used.

BERLAND, L. *Les Arachnides*. Lechevalier, Paris, 1932

BONNET, P. *Bibliographia Araneorum*. Toulouse. 3 volumes, 1945–1961

BRISTOWE, W.S. *The World of Spiders*. Collins, 1958

CHILD, JOHN. *Australian Spiders*. Periwinkle Press, Gladesville, 1965

COMSTOCK, J.H. *The Spider Book*. Doubleday and Co. New York, 1912, revised edition 1940, W.J. Gertsch

DALMAS, RAYMOND DE. *Araignées de Nouvelle Zélande*. Ann. Soc. ent. France 86 pp. 431–436

FORSTER, R.R. The New Zealand Harvestmen (Suborder Laniatores). *Canterbury Museum Bulletin No. 2*, Christchurch, 1954

FORSTER, R.R. A Key to the New Zealand Harvestmen. *Tuatara*, 10 (3): 129–137. 11(1): 28–40

FORSTER, R.R. The Spiders of New Zealand. Part 1. *Otago Museum Bulletin No. 1*, Dunedin, 1967

GERTSCH, W.J. *American Spiders*. Van Nostrand, 1949

KASTON, B.J. *How to know the Spiders*. W.C. Browne, Dubuque, Iowa

LOCKET, G.H. and MILLIDGE, A.F. *British Spiders*, 2 volumes. The Ray Society, 1951, 1953

MAIN, BARBARA. *Spiders of Australia*. The Jacaranda Press.

New Zealand Spiders

McKeown, K.C. *Australian Spiders. Their lives and habits.* Angus and Robertson, 1952

Parrott, A.W. A Systematic Catalogue of New Zealand Spiders. *Records of the Canterbury Museum.* 5:51–92. Christchurch, 1946

Roewer, C.F. *Katalog der Araneae 1758–1942.* 2 volumes Bremen, 1942–1954

Savory, T.H. *The Biology of Spiders.* Sidgwick and Jackson, London, 1928

Savory, T.H. *The Arachnida.* Edward Arnold and Co. London, 1935

Savory, T.H. *Spiders, Men and Scorpions.* University of London Press, London, 1961

Savory, T.H. *Spiders and other Arachnids.* The English Universities Press, 1964

Yaginuma, T. *Spiders of Japan in Colour.* Hoikusha, Osaka, 1960

Index

Numerals in **bold type** refer to illustrations. Those in colour are given the page reference nearest to, and preceding, the plate.